Lecture Notes in Computer Science 11977

More information about this series at http://www.springer.com/series/7412

Quanzheng Li · Richard Leahy · Bin Dong ·
Xiang Li (Eds.)

Multiscale Multimodal Medical Imaging

First International Workshop, MMMI 2019
Held in Conjunction with MICCAI 2019
Shenzhen, China, October 13, 2019
Proceedings

 Springer

Editors
Quanzheng Li (iD)
Harvard Medical School
Boston, MA, USA

Richard Leahy
University of Southern California
Los Angeles, CA, USA

Bin Dong
Peking University
Beijing, China

Xiang Li (iD)
Harvard Medical School
Boston, MA, USA

ISSN 0302-9743 ISSN 1611-3349 (electronic)
Lecture Notes in Computer Science
ISBN 978-3-030-37968-1 ISBN 978-3-030-37969-8 (eBook)
https://doi.org/10.1007/978-3-030-37969-8

LNCS Sublibrary: SL6 – Image Processing, Computer Vision, Pattern Recognition, and Graphics

This Springer imprint is published by the registered company Springer Nature Switzerland AG
The registered company address is: Gewerbestrasse 11, 6330 Cham, Switzerland

Preface

In 2019, the First International Workshop on Multiscale Multimodal Medical Imaging (MMMI 2019), a workshop held in conjunction with International Conference on Medical Image Computing and Computer-Assisted Intervention (MICCAI 2019), took place in Shenzhen, China. It was organized by the Massachusetts General Hospital and Harvard Medical School, the University of Southern California, and Peking University.

In the field of medical imaging, the use of more than one imaging modality (i.e. multimodal) or analyzing images at a different scale (i.e. multiscale) on the same target has become a growing field, as more advanced techniques and devices have become available. Various analyzes using multiscale/multimodal medical imaging and computer-aided detection systems have been developed, with the premise that additional modalities/scales can encompass abundant information which is different and complementary to each other. Facing the growing amount of data available from multiscale/multimodal medical imaging facilities, a variety of new methods for the image analysis have been developed so far. The MMMI workshop aims to move forward the state of the art in multiscale/multimodal medical imaging, including both algorithm development, implementation of methodology, and experimental studies. The workshop also aims to facilitate more interactions between researchers in the field of medical image analysis and the field of machine learning, especially in machine learning methods for data fusion and multisource learning.

The MMMI workshop took place at the InterContinental Hotel Shenzhen in Shenzhen, China, on October 13, 2019. It attracted more than 50 registered attendees from international communities of computer scientists, imaging physics, radiologists, and clinical physicians, who presented works covering a wide range of medical imaging modalities and applications. Novel techniques and insights for multiscale/multimodal medical images analysis, as well as empirical studies involving the application of multiscale/multimodal imaging for clinical use were presented. MMMI 2019 received a total of 18 submissions, which were reviewed by 29 independent reviewers. All submissions underwent a double-blind peer-review process, with each submission being reviewed by at least two independent reviewers and one Program Committee member. Based on the review scores and comments, 13 papers were accepted for presentation at the workshop and for inclusion in this Springer LNCS volume.

We greatly appreciate all the author's contributions to this workshop. We would like to thank all the Program Committee members for handling the submissions with professional judgements and constructive comments. We also thank our sponsors for the financial supports of the Best Paper Awards and Student Paper Awards.

October 2019

Quanzheng Li
Richard Leahy
Bin Dong
Xiang Li

Organization

Program Committee Chairs

Quanzheng Li   Massachusetts General Hospital, USA
Richard Leahy   University of Southern California, USA
Bin Dong   Peking University, China
Xiang Li   Massachusetts General Hospital, USA

Program Committee

Dufan Wu   Massachusetts General Hospital and Harvard Medical
                        School, USA
Jian Li   University of Southern California, USA
Jie Zhang   Arizona State University, USA
Jinglei Lv   QIMR Berghofer Medical Research Institute, Australia
Kyungsang Kim   Massachusetts General Hospital and Harvard Medical
                        School, USA
Shijie Zhao   Northwestern Polytechnical University, China
Shu Zhang   Stony Brook University, USA
Xi Jiang   University of Electronic Science and Technology
                        of China, China
Xiao-Yun Zhou   Imperial College London, UK

Contents

Multi-modal Image Prediction via Spatial Hybrid U-Net

Akib Zaman[1(⊠)], Lu Zhang[1], Jingwen Yan[2], and Dajiang Zhu[1]

[1] The University of Texas at Arlington, Arlington, USA
akbzmn@gmail.com
[2] Indiana University-Purdue University Indianapolis, Indianapolis, USA

Abstract. Cortical folding patterns and white matter connectivity together compose the structural organization of human brain. Gray matter and gyrification describe the geometric characteristic of cortical surface and the wiring of white matter represents the structural pathway inside the brain. Many studies suggest that there exists a close relationship between gray matter and white matter. However, given the widely existing variability and complexity of brain structures, it is still largely unknown to what extent white matter wiring can influence gray matter and folding patterns. As an attempt to discover the potential relationship between gray matter and white matter, in this work we developed a novel spatial hybrid U-Net framework for multi-modal image prediction: we are aiming to predict T1-weighted Magnetic Resonance Imaging (MRI) based on Diffusion Tensor Imaging (DTI) data. Specifically, when predicting local intensity for T1 data, we constructed a hybrid model to integrate both local tensor information and the FA (Fractional Anisotropy) measure from remote brain regions connected by DTI derived fibers. To alleviate computation effort and reduce memory consumption, we proposed a multi-stage 2D training scheme instead of using 3D convolution neural network. Our results showed 80% accuracy for prediction and the reconstructed cortical surface using predicted T1 data is highly consistent to the original T1 derived surface. We envision that the proposed method can not only lay down a foundation for multi-modality inference, but also bring new insights to understand brain structure as well.

Keywords: Brain structure · Multi-modality · U-Net

1 Introduction

Human brains display significant inter-individual variation in cortical structures. Folding pattern of the cerebral cortex (gray matter - GM) and white matter (WM) connectivity are two aspects of brain structure. These two characteristics together compose the structural organization of human brain. That is, gray matter and gyrification depict the geometric shape of cortical surface, e.g., gyri and sulci, and white matter wiring provides the white matter pathway inside the cortex. Many studies suggest that there exists a close relationship between gray matter and white matter, such as a universal scaling law between GM and WM [1]. There are two divergent ideas in the field: axonal pushing and pulling [2] theory. A recent study further proved that during cortical gyrification, gyral regions with higher concentrations of growing axonal fibers tend to form 3-hinge

© Springer Nature Switzerland AG 2020
Q. Li et al. (Eds.): MMMI 2019, LNCS 11977, pp. 1–9, 2020.
https://doi.org/10.1007/978-3-030-37969-8_1

gyri [3]. All the above studies suggest that there exists a close relationship between gray matter and white matter. However, given the existing variability and complexity of brain structures, it is still largely unknown to what extent white matter wiring can influence gray matter and folding patterns.

Fortunately, the advancement of MRI and DTI provides non-invasive ways to examine cortical folding patterns and white matter related measures (e.g. FA maps and DTI derived fibers). As an attempt to discover the potential relationship between gray matter and white matter, we developed a novel spatial hybrid U-Net framework for multi-modal image prediction: we aim to predict T1-weighted MRI based on DTI data. The motivation is that if white matter influences cortical folding (gray matter) via either pulling/pushing or their combined effects, we should be able to infer gray matter with the information of white matter. Specifically, when predicting local intensity for T1-weighted image data, we constructed a hybrid model to integrate both local tensor information and the FA measures from remote brain regions connected by DTI derived fibers. The reason for integrating local and remote white matter knowledge as a hybrid model is to examine if structural connectivity will also contribute to local gray matter properties. To alleviate computation effort and reduce memory consumption, we proposed a multi-stage 2D training scheme instead of using 3D convolution neural network (CNN). Using Human Connectome Project [4] data set as a test bed, our results showed 80% accuracy for T1-weighted image prediction with DTI data. Here, the accuracy is defined as the ratio of predicted intensity to the original one. We also examined the reconstructed cortical surfaces with the predicted T1 data and it displayed high similarity to the surfaces generated from the original T1 image.

2   Method

2.1   Data Acquisition and Pre-processing

The data used in this work is acquired from the WU-Minn Human Connectome Project (HCP) consortium S1200 Release. We use T1-weighted MRI and DTI. For T1, TR = 2.4 s, TE = 2.14 ms, voxel size is 0.7 mm isotropic. For DTI, TR = 5.520 s, TE = 89.5 ms, slice thickness is 1.25 mm. The diffusion weighted data consists of 3 shells of b = 1000, 2000, and 3000 s/mm^2 with an approximately equal number of acquisitions on each shell within each run. In this work, we only used 90 b = 1000 volumes and 1 b = 0 volume.

The data acquired is then pre-processed through a series of steps using the FMRIB Software Library (FSL) [5, 6]. Finally, the T1-weighted images are registered to their respective DTI b = 0 images using FMRIB's Linear Image Registration Tool. [7, 8] This step allows us to register the DTI and FA images in the same space as the T1 image.

2.2   Extracting Features from Data

To alleviate computation effort and reduce memory consumption, we proposed a multi-stage 2D training scheme instead of using 3D CNN. That is, each data sample for our model training and prediction is generated based on a single 2D slice. Figure 1 demonstrates the construction of our input. The features contained in each sample comprise two parts: the first part comes from local structural information which includes the gradients of ninety b = 1000 volumes and one b = 0 volume. The second part of the input represents the remote structural information. For example, for each voxel considered, we can compute its connected voxels (the colored voxels in Fig. 1) in remote regions by examining if there are fibers passing through both of them. For the connected voxels, we acquire their FA measures and concatenate them with the local information as the entire input. Because of computation constrains, we considered 6 remote regions for each slice. Different features are treated as different channels. Totally we have 97 channels for each 2D slice. The motivation to integrate both local and remote structural information together for model training and prediction is to examine if the remote brain regions can influence the local brain structures via WM structural connectivity.

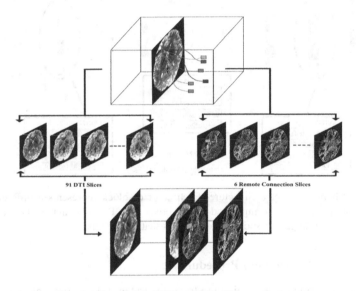

Fig. 1. Combining DTI and FA slices as features

2.3   Transfer Learning Using a Spatial Hybrid U-Net Model

The deep neural network architecture adopted in this work is similar to the U-Net model [9]. This model can be divided into two major parts with 5 blocks on the left (down-sampling) and 5 blocks on the right (up-sampling) of a central block. The first block of the model represents the input layer. This is a collection of 2D image slices from the 3D volume we generated from the DTI and FA images. This is followed by four blocks of similar structure (down sampling) - two 3 × 3 2D convolutions with a rectified linear unit (ReLU), followed by a 2 × 2 max pooling operation. Batch normalization and a dropout of 25% is used after each max pooling layer. Before we begin the up-sampling blocks, we apply two 3 × 3 convolutions with ReLU followed by batch normalization and a dropout of 50% nodes. The four blocks of the up-sampling part consist of a concatenate operation to merge the layers in the same spatial order. This is the key feature of the U-Net architecture.

This is followed by two similar 3×3 2D convolution operations and batch normalization. The final up sampling block is passed through a 1×1 2D convolution with a single filter to regress the output. Note that even though the input of the proposed model is based on 2D slices, it contains the 3rd dimension of 97 channels which integrate both local and spatially remote structural information. Therefore, we named it as spatial hybrid U-Net. Figure 2 visualizes the overall structure we used in this adaptation of the U-Net.

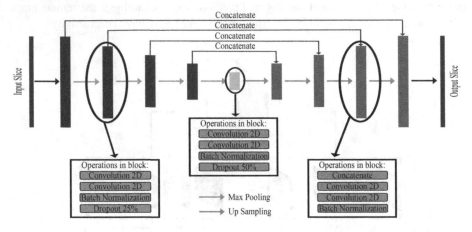

Fig. 2. Spatial hybrid U-Net architecture. Each color of block represents a different type of combination of operations. The input and output boxes are 2D slices images. One block from each type has been enlarged to show the operations that it contains.

2.4   Training and Prediction Procedure

In this work, we proposed a multi-stage training strategy: the entire training process is divided into three stages corresponding to three image directions. First, we train the model from the sagittal plane. For this, we use the image slices pre-processed with the sagittal direction and feed into our spatial hybrid U-net model. These image slices are from 15 subjects (subject 1 to 15) and each slice is treated as an independent training sample. This part of the training helps us achieve a baseline for the weight initialization, which will be used to train the model with images from other planes. The images from the axial plane are taken from a different set of 15 subjects (subject 16 to 30) and the model is continued to be trained from the axial direction. Finally, we train the model from the coronal direction using coronal slices from another set of 15 subjects (subject 31 to 45). The reason that we did not conduct the training using three directions simultaneously is that there exists significant difference of the training performance for different directions. Essentially, we trained the model from a direction with the best prediction performance and used that as pre-knowledge to train the next one. This is similar to transfer learning that transfers the knowledge gained from one plane to a different plane. Figure 3 shows the overview of our training process.

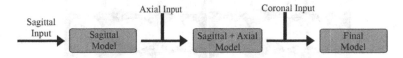

Fig. 3. Overview of training process

The optimizer used to train the model is the stochastic gradient-based Adam optimizer. It updates the model parameters with the rule:

$$\Omega^{t+1} = \Omega^t - \gamma \cdot \frac{\widehat{m_a^t}}{\sqrt{\widehat{m_b^t}} + v} \tag{1}$$

The Adam optimizer computes network weights and biases Ω^{t+1} for step $t + 1$ using adaptive momentum estimation technique, which enables it to compute parameters using variable learning rate (γ) instead of a fixed constant learning rate. This enables the optimizer to calculate gradient steps more precisely enabling it to converge faster. It uses first and second exponential decay rates β_a and β_b to compute first and second momentum estimates $\widehat{m_a^t}$ and $\widehat{m_b^t}$. These values are used to compute new weights and biases for the network at $t + 1$ step. The parameter v is a regularization parameter used to avoid division by zero cases.

We use Mean Squared Error to penalize the difference in the predicted value. The loss is calculated by the following formula:

$$f(\Omega^t) = \frac{1}{N} \sum_{i=1}^{N} (t_i - p_i)^2 \tag{2}$$

Here, t_i and p_i are true and predicted values of the deep neural network and N is a normalizing constant defined by the number of subjects (number of training samples).

3   Results

3.1   Predicting T1-Weighted Images from DTI and FA Images

Given the 90 gradients extracted from 90 b = 1000 s/mm^2 DWI volumes (local information) as well as FA measures of voxels connected (remote information) for each voxel (Sect. 2.2), we conducted T1 prediction model training with spatial hybrid U-Net

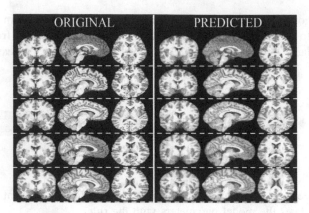

Fig. 4. Comparison of predicted T1-weighted image and original T1 data. We used 45 subjects for prediction model training (15 subjects for each stage). Here, we randomly select 5 subjects and show their prediction results (right) with the original T1 data (left) as well

frame-work proposed in Sects. 2.3 and 2.4. In this work, we perform the prediction (50 subjects) from all three directions and the final result is averaged. We randomly select 5 subjects and show their original T1 data as well as prediction results in Fig. 4. We can see that both overall structures and detailed spatial patterns of GM/WM are highly consistent between our prediction results and the original T1 data. The related quantitative results are shown in Sect. 3.3.

3.2   Predicting Cortical Surfaces with Predicted T1-Weighted Images

For the 50 predicted T1 images using our DTI prediction model, we conducted the cortical surface reconstruction as regular T1 images. Each sub-figure in Fig. 5 shows the reconstructed cortical surfaces using the original T1 image (top) and our predicted one (bottom). By visual examination we can see that the overall folding patterns are highly consistent between the original T1 derived cortical surfaces and the ones based on DTI prediction results. The color encodes the prediction error and the regions with red represent high prediction error. An interesting observation is that the temporal lobe (red circles) and dorsal regions (red arrows) tend to have higher prediction error compared to other brain regions. This might be due to the more complex structure of the corresponding regions. Importantly, we can accurately predict T1 image and generate cortical surface only using DTI data, which was considered a different imaging modality.

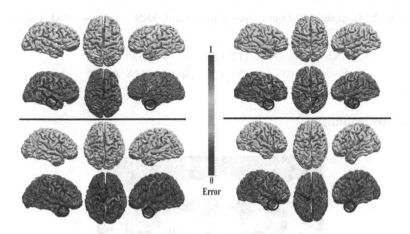

Fig. 5. Examples of cortical surface reconstruction using original and predicted T1 images (Color figure online)

3.3 Comparison with Model Without FA Slices in the Training Data

In our paper, we used Mean Absolute Error (MAE) to measure the quality of predicted T1-weighted images. MAE is a measure of difference between two continuous variables. The Mean Absolute Error is given by:

$$\text{MAE} = \frac{\sum_{i=1}^{n} |a_i - b_i|}{n} \tag{3}$$

where a_i and b_i are the original T1 data and predicted T1 data, respectively.

Our comparison results can be seen in Table 1. 5 predicted subjects were randomly chosen and compared with their original registered T1-weighted image. It is evident that including remote features (FA) has a positive effect on the prediction model.

Table 1. Comparison of our proposed DTI + FA method with only DTI method (MAE values)

Subject	1	2	3	4	5	Average
DTI + FA method	0.073	0.072	0.074	0.057	0.073	**0.0714**
Only DTI method	0.081	0.079	0.082	0.069	0.084	0.0790

3.4 Comparison with k-Nearest Neighbor (kNN) Regression

We compared the results of our model with kNN Regression (n = 5) and it is evident from the results that our proposed model performs better. The MAE values have been compared to show the performance improvement that our model achieves (Table 2).

Table 2. Comparison of our proposed model with kNN regression (MAE values)

Subject	1	2	3	4	5	Average
kNN	0.120	0.112	0.115	0.109	0.130	0.1172
Proposed	0.073	0.054	0.057	0.063	0.068	**0.0630**

3.5   Reproducibility

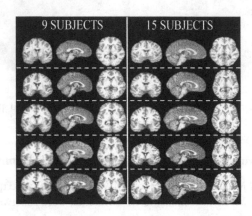

Fig. 6. Prediction using 9 subjects (left) and 15 subjects (right)

We examined the robustness of our model by using the same training procedure for 9 subjects (3 from each plane) and 15 subjects (5 from each plane). The results were satisfactory, as can be seen from Fig. 6 above. Using higher number of subjects yields higher accuracy, as expected. However, our model proves to be robust in terms of the T1-weighted predictions even with very few subjects used for training.

4   Conclusion

In this work, we developed a novel spatial hybrid U-Net model to predict T1 image from DTI data. During this process, we used both the local structural information and also FA measures from remote regions connected by DTI derived fibers. We also proposed a multi-stage training scheme to achieve a more reliable prediction performance. Our results showed 80% accuracy for prediction and the reconstructed cortical surface using predicted T1 data is highly consistent to the original T1 derived surface. We envision that the proposed method can not only lay down a foundation for multi-modality inference, but also bring new insights to understand brain structure as well.

References

1. Zhang, K., Sejnowski, T.J.: A universal scaling law between gray matter and white matter of cerebral cortex. Proc. Natl. Acad. Sci. U.S.A. **97**(10), 5621–5626 (2000)
2. Razavi, M.J., et al.: Radial structure scaffolds convolution patterns of developing cerebral cortex. Front. Comput. Neurosci. **11**, 76 (2017)
3. Ge, F., et al.: Denser growing fiber connections induce 3-hinge gyral folding. Cereb. Cortex **28**(3), 1064–1075 (2017)
4. Van Essen, D.C., et al.: The WU-Minn human connectome project: an overview. NeuroImage **80**, 62–79 (2013)
5. Jenkinson, M., Beckmann, C.F., Behrens, T.E., Woolrich, M.W., Smith, S.M.: FSL. NeuroImage **62**, 782–790 (2012)
6. Smith, S.M., et al.: Advances in functional and structural MR image analysis and implementation as FSL. NeuroImage **23**(S1), 208–219 (2004)
7. Jenkinson, M., Smith, S.M.: A global optimisation method for robust affine registration of brain images. Med. Image Anal. **5**(2), 143–156 (2001)
8. Jenkinson, M., Bannister, P.R., Brady, J.M., Smith, S.M.: Improved optimisation for the robust and accurate linear registration and motion correction of brain images. NeuroImage **17**(2), 825–841 (2002)
9. Ronneberger, O., Fischer, P., Brox, T.: U-Net: convolutional networks for biomedical image segmentation. In: Navab, N., Hornegger, J., Wells, W., Frangi, A. (eds.) MICCAI 2015. LNCS, vol. 9351, pp. 234–241. Springer, Cham (2015). https://doi.org/10.1007/978-3-319-24574-4_28

Automatic Segmentation of Liver CT Image Based on Dense Pyramid Network

Hongli Xu[1], Binhua Wang[1(✉)], Wanguo Xue[1], Yao Zhang[2,3],
Cheng Zhong[4], Yongliang Chen[5(✉)], and Jianfeng Leng[6]

[1] Medical Big Data Center, The First Medical Center,
Chinese PLA General Hospital, Beijing, China
xuhongli@301hospital.com.cn, 7539wbhwbh@163.com
[2] Institute of Computing Technology, Chinese Academy of Sciences,
Beijing, China
[3] University of Chinese Academy of Sciences, Beijing, China
[4] Lenovo Research, Lenovo Ltd., Beijing, China
[5] Department of Hepatobiliary Surgery, The First Medical Center,
Chinese PLA General Hospital, Beijing, China
chenyl301@163.com
[6] Department of Hepatobiliary Surgery, Peking University Shougang Hospital,
Beijing, China

Abstract. In the field of computer-aided diagnosis (CAD) and treatment evaluation system on hepatic disease diagnosis, the automatic segmentation of the liver from CT volume is the most basic step. It is a difficult task because the shape of liver could be changed by liver tumor, and the intensity of liver is similar to that of other adjacent tissues. In this paper, we proposed a framework based on the U-net architecture, called dense pyramid network to segment the liver from CT images automatically. The main contribution of our network is that, multiple feature maps from the previous level of hierarchy are combined as the input of each layer in the encoding part. This removes the loss of context information between different layers. The model is trained on practical enhanced CT scans, which are gained from People's Liberation Army General Hospital (PLA). Experimental results demonstrate that our model can effectively improve the segmentation performance of liver, no matter the different shapes between livers. In the experiment, the Dice score of liver segmentation in the arterial phase, venous phase, and delay phase by dense pyramid network was about 95.97%, 96.22%, and 96.16%, respectively, which shows that our model is more suitable for multi-phase liver segmentation.

Keywords: Deep learning · Liver segmentation · U-Net · Dense pyramid network

1 Introduction

Liver cancer is a common malignant tumor in the world. In 2013, the incidence and mortality of liver cancer in China were higher than the global average [1, 2], the early diagnosis and treatment of liver cancer has always been received highly attention.

© Springer Nature Switzerland AG 2020
Q. Li et al. (Eds.): MMMI 2019, LNCS 11977, pp. 10–16, 2020.
https://doi.org/10.1007/978-3-030-37969-8_2

In medical imaging technology, CT has become an important reference because of its high spatial resolution, high robustness and short acquisition time [3–5]. Currently, multi-phase CT images are widely used to diagnose hepatic disease, but when clinical experts diagnose the type of liver cancer, they usually need roll back and forth in different phases, even zoom in and out the CT images; besides manual annotation on thousands of CT slices is time-consuming [6], therefore, Accurate liver automatic segmentation to determine the size of the liver and the need for tumor location is an urgent problem. Accurate liver segmentation on CT images is a challenging task because the intensity is similar between the liver and other organs around it, and the size and contour of the liver can be affected by liver tumors or other diseases. To segment the liver issue based on contrast-enhanced CT images, the traditional image segmentation algorithm is to extract image features based on the distribution of HU values of CT images, such as level set [7, 8] based methods and graph cut [9, 10] based methods. However, the disadvantage is that the parameters of the model need to be initialized according to the doctor's prior knowledge, which is greatly influenced by subjective factors.

In recent years, deep learning has become popular in the field of medical imaging. Litjens [11, 12] et al. summarized the application of deep learning methods in the field of medical image processing. The U-net architecture proposed by Ronneberger [13] et al. apparently becomes a main method on medical images segmentation, and makes an impressive improvement by its elegant structure of skip connections. It consists of an end-to-end contraction path and expansion path, which follows a basic encoder-decoder architecture, the contraction path is consecutive of convolution layer and max-pooling layer, which extract more features pass to the next layer. Downsampling is performed by a convolution-pooling operation during the encoding process to obtain a low-resolution feature map; the expansion path is to upsample the feature map to recover the size of segmentation map, and Skip-connection brings feature maps in the encoder to the decoder of the same level so that the segmentation mask can be synthesized. It has the following shortcomings: the context information of higher resolution layers can't be propagated to the deep convolutional layer; when the number of layers is increased, the gradient return path is longer, which is not conducive to the convergence of the model.

In our work, we proposed a framework based on U-net, called dense pyramidal feature network to segment multi-phase CT images of liver tissue automatically, which can reduce users' bias and segmentation time, meanwhile, it's the foundation for liver tumor segmentation and surgical planning.

2 Method

An overview of our proposed dense pyramid architecture is shown in Fig. 1. Firstly, the enhanced CT images are pre-processed using HU clip, then the images are fed into our segmentation network, which is inspired by the experience of U-net. We will introduce the dataset we used in our work, then the detailed dense pyramid architecture will be given.

2.1 Dataset

The model is trained and tested on practical abdominal enhanced DICOM images chosen from People's Liberation Army General Hospital (PLA), the database is composed of 15 males and 5 females, and each CT volume include three phases: arterial phase, venous phase, delay phase, among them, there are 16 CT volumes contain hepatic tumors or other related illness.

2.2 Dense Pyramid Network

Our model is trained based on dense pyramid network, which is optimized on the U-net, and the schematic is shown in Fig. 1.

Compared with traditional fully convolutional neural network, which is shown in Fig. 2, the dense pyramid network is composed of end-to-end network architecture, multi-scale pyramid features and dense connection structure. The end-to-end architecture is a basic encoder-decoder architecture, and the loss function is weighted by cross entropy [14], which is defined as:

$$\text{loss} = -\frac{1}{N}\sum_{i=1}^{N} w_i \left[\widehat{P_i} log P_i + \left(1 - \widehat{P_i} \right) \log(1 - P_i) \right] \tag{1}$$

Where P_i denotes the probability of pixel i belonging to the foreground, $\widehat{P_i}$ represents the ground truth, and w_i is the weighted factor. The multi-scale pyramid features is shown in Fig. 1, it gains different feature maps by max-pooling layers, finally, the dense connection, multiple feature maps from the previous level of hierarchy are combined as the input of each layer in the encoding part, this removes the loss of context information between different layers.

The model is implemented with Tensorflow library. We train the network from scratch with a Gaussian random initializer ($\mu = 0$, $\sigma = 0.01$). The Adam optimizer with an initial learning rate of 0.0001 is used for parameters updating. By considering the sizes of background and liver, we set the weights of the cross entropy loss to be 1 and 16 respectively. The followings are the specific experimental steps of this study:

Firstly, before using the CT images, we replaced the privacy information of the patients which is included in the DICOM header file, secondly, we manually annotated the enhanced CT images on the direction of Transverse plane which are guided by experts, then, the Hounsfield unit values of each CT image are ranged in [−75, 175], finally, this study used a 4-fold cross-validation method to obtain liver segmentation results, and connected domain analysis was performed on the segmentation results, and the largest connected domain was retained as the final liver segmentation result.

2.3 Evaluation Metrics

The liver segmentation result is measured according to commonly used Dice score [15], which is defined as:

$$\text{DICE}(A, B) = \frac{2|A \cap B|}{|A| + |B|} \tag{2}$$

We refer to foreground in the ground truth as A, and the predicted foreground as B, and the higher the Dice score is, the better the liver segmentation result is.

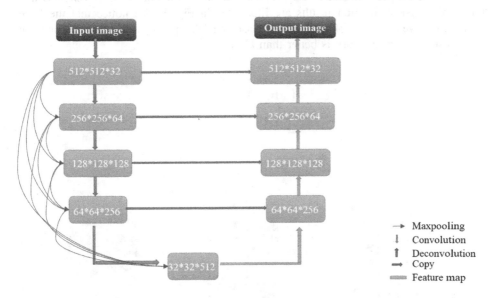

Fig. 1. Overview of the proposed dense pyramid network

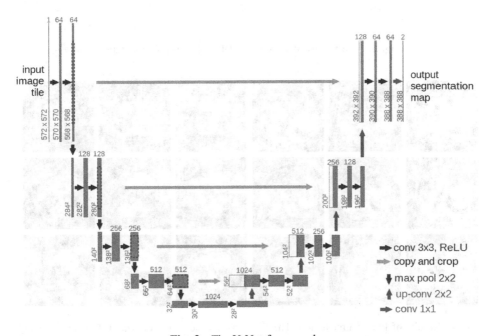

Fig. 2. The U-Net framework

3　Result and Discussion

We evaluate the performance of our proposed dense pyramid network as well as U-net, the quantitative evaluation results are illustrated in Fig. 3. We notice that the Dice scores improve in arterial phase, venous phase and delay phase, which demonstrate that the dense pyramid network is better than U-net.

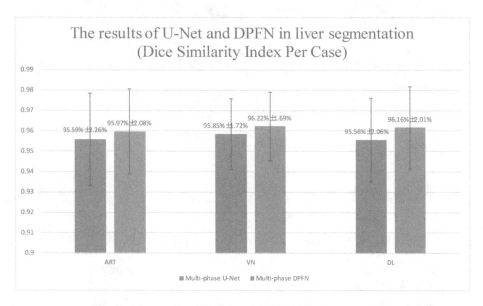

Fig. 3. The results of U-Net and DPFN in liver segmentation

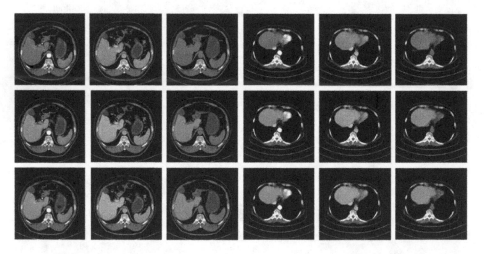

Fig. 4. Results of liver segmentation. The first row represents the original CT images, second row represents ground truth, and third row represents results of liver segmentation with dense pyramid network (the liver is labelled in red). The first three columns are in order of ART, VN, and DL phase respectively. The last three columns are in the same order. (Color figure online)

In Fig. 4, we show some representative segmentation examples obtained with the dense pyramid network. According to the results, we conclude that the dense pyramid network we proposed can detect the most liver region, and adapt to different liver forms no matter the shapes of the livers among different CT volumes are various.

Besides, the enhanced CT images we used are composed of arterial phase, venous phase and delay phase, and different phases have different characteristics on image, which are shown in Fig. 5, we can achieve an effective segmentation result of multiple phases of liver images using only one segmentation network.

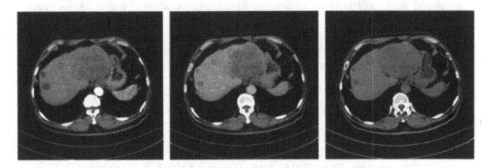

Fig. 5. CT images of liver in multi-phases

4 Conclusion and Future Work

In this paper, we design and proposed a model, called dense pyramid network to segment liver from abdominal enhanced CT scans automatically. Within our model, multiple feature maps from the previous level of hierarchy are combined as the input of each layer in the encoding part to combine the detail texture and the contour information, which can solve the multi-scale problem in image segmentation to some extent, and improve the performance of liver segmentation of CT scans. In the future, we would like to design some efficient algorithm to achieve the segmentation of the liver tumor on the basis of liver segmentation, so as to achieve the purpose of computer-aided diagnosis and evaluation of treatment plans.

References

1. Chen, W., Zheng, R., Baade, P.D., et al.: Cancer statistics in China, 2015. CA Cancer J. Clin. **66**(2), 115–132 (2016)
2. Bray, F., Ferlay, J., Soerjomataram, I., et al.: Global cancer statistics 2018: GLOBOCAN estimates of incidence and mortality worldwide for 36 cancers in 185 countries. CA Cancer J. Clin. **68**(6), 394–424 (2018)
3. Nakayama, Y., Li, Q., Katsuragawa, S., et al.: Automated hepatic volumetry for living related liver transplantation at multisection CT. Radiology **240**(3), 743–748 (2006)
4. Masutani, Y., Uozumi, K., Akahane, M., et al.: Liver CT image processing: a short introduction of the technical elements. Eur. J. Radiol. **58**(2), 246–251 (2006)

5. Campadelli, P., Casiraghi, E., Esposito, A.: Liver segmentation from computed tomography scans: a survey and a new algorithm. Artif. Intell. Med. **45**(2–3), 185–196 (2009)
6. Ecabert, O., Peters, J., Schramm, H., et al.: Automatic model-based segmentation of the heart in CT images. IEEE Trans. Med. Imaging MI **27**(9), 1189–1201 (2008)
7. Cremers, D., Rousson, M., Deriche, R.: A review of statistical approaches to level set segmentation: integrating color, texture, motion and shape. Int. J. Comput. Vis. **72**(2), 195–215 (2007)
8. Lecellier, F., Jehan-Besson, S., Fadili, J.: Statistical region-based active contours for segmentation: an overview. IRBM **35**(1), 3–10 (2014)
9. Afifi, A., Nakaguchi, T.: Liver segmentation approach using graph cuts and iteratively estimated shape and intensity constrains. Med. Image Comput. Comput. Assist. Interv. **15**(2), 395–403 (2012)
10. Huang, Q., Ding, H., Wang, X., et al.: Fully automatic liver segmentation in CT images using modified graph cuts and feature detection. Comput. Biol. Med. **95**, 198 (2018)
11. Litjens, G., Kooi, T., Bejnordi, B.E., et al.: A survey on deep learning in medical image analysis. Med. Image **42**(9), 60–88 (2017)
12. Shen, D., Wu, G., Suk, H.I.: Deep learning in medical image analysis. Ann. Rev. Biomed. Eng. **19**(1), 221–248 (2017). https://doi.org/10.1146/annurev-bioeng-071516-044442
13. Ronneberger, O., Fischer, P., Brox, T.: U-Net: convolutional networks for biomedical image segmentation. In: Navab, N., Hornegger, J., Wells, W., Frangi, A. (eds.) Medical Image Computing and Computer-Assisted Intervention – MICCAI 2015. LNCS, vol. 9351, pp. 234–241. Springer, Cham (2015). https://doi.org/10.1007/978-3-319-24574-4_28
14. Christ, P.F., Ettlinger, F., Grün, F., et al.: Automatic liver and tumor segmentation of CT and MRI volumes using cascaded fully convolutional neural networks (2017)
15. Heimann, T., van Ginneken, B., et al.: Comparison and evaluation of methods for liver segmentation from CT datasets. IEEE Trans. Med. Imaging **28**(8), 1251–1265 (2009)

OctopusNet: A Deep Learning Segmentation Network for Multi-modal Medical Images

Yu Chen[1,2], Jiawei Chen[2], Dong Wei[2], Yuexiang Li[2(✉)], and Yefeng Zheng[2]

[1] Nanjing University, Nanjing, China
[2] YouTu Lab, Tencent, Shenzhen, China
vicyxli@tencent.com

Abstract. Deep learning models, such as the fully convolutional network (FCN), have been widely used in 3D biomedical segmentation and achieved state-of-the-art performance. Multiple modalities are often used for disease diagnosis and quantification. Two approaches are widely used in the literature to fuse multiple modalities in the segmentation networks: early-fusion (which stacks multiple modalities as different input channels) and late-fusion (which fuses the segmentation results from different modalities at the very end). These fusion methods easily suffer from the cross-modal interference caused by the input modalities which have wide variations. To address the problem, we propose a novel deep learning architecture, namely OctopusNet, to better leverage and fuse the information contained in multi-modalities. The proposed framework employs a separate encoder for each modality for feature extraction and exploits a hyper-fusion decoder to fuse the extracted features while avoiding feature explosion. We evaluate the proposed OctopusNet on two publicly available datasets, i.e. ISLES-2018 and MRBrainS-2013. The experimental results show that our framework outperforms the commonly-used feature fusion approaches and yields the state-of-the-art segmentation accuracy.

Keywords: Medical image segmentation · Deep learning · Multi-modal images

1 Introduction

Recent years have witnessed the rapid development of deep learning technique. Deep learning models have been widely used for medical image segmentation and achieved impressive performance [1–3]. Compared with natural images, medical images, e.g. computed tomography (CT) and magnetic resonance imaging (MRI), often have a lot of scanning protocols in its toolbox and each protocol may reveal a different property (often complementary to other protocols) of the underlying tissue. For examples, to assess ischemic stroke lesion, three modalities using perfusion imaging are commonly captured, i.e. cerebral blood volume

---

This work was done when Yu Chen was an intern at YouTu Lab.

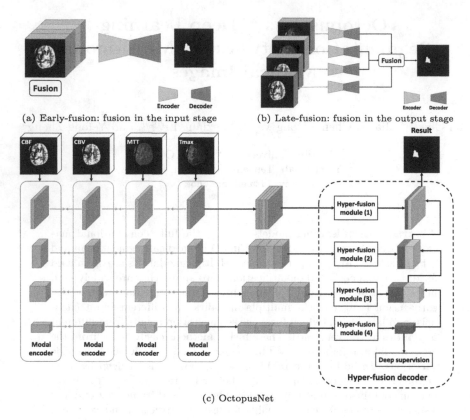

(a) Early-fusion: fusion in the input stage

(b) Late-fusion: fusion in the output stage

(c) OctopusNet

Fig. 1. Different fusion approaches, including early-fusion (a), late-fusion (b) and the architecture of our OctopusNet (c).(Color figure online)

(CBV), cerebral blood flow (CBF), and time to peak of the residue function (Tmax). Those modal images may contain different clinical interpretation.

To exploit the multi-modal medical data, two fusion approaches are widely-used by current deep learning networks, i.e. stacking multiple modalities as different input channels (early-fusion, Fig. 1(a)) [1,2,4] and fusing the outputs of networks from different modalities (late-fusion, Fig. 1(b)) [3,5]. Neither fusion approach is optimal in using the complementary information from multiple input modalities. Take the perfusion CT for ischemic stroke lesion segmentation as an example. As shown in Fig. 1(c), four modalities, i.e. CBV, CBF and MTT, and Tmax, are captured. It can be observed from the four modalities that the lesion area in CBV and CBF is darker compared to the normal area, while it is lighter in the modalities of MTT and Tmax. Consequently, the information in different modalities may be wrongly fused if we simply adopt the early-fusion approach. On the other hand, although the late fusion approach adopts the separate encoder-decoder for each modality, the whole network is computational expensive and difficult to converge.

In this paper, we propose a novel segmentation network, namely Octopus-Net, which effectively leverages the information contained in multi-modal medical images. Instead of fusing multi-modal images at the input stage, we exploit an individual encoder for each modality and fuse the feature maps generated by middle stages of the network, which specifically extracts features from each modality and explicitly considers the correlations between different modalities. As the modalities are separately encoded, the proposed OctopusNet adopts a novel feature fusion module, namely a hyper-fusion decoder, to merge the feature maps and avoid feature explosion. Extensive comparison experiments are conducted on multi-modal datasets. The results demonstrate the outstanding segmentation performance of the proposed OctopusNet.

2   OctopusNet

In this section, we introduce the detailed information of our OctopusNet. The framework of our OctopusNet is shown in Fig. 1(c). The colored cubes refer to the feature maps generated at different stages of the framework. Our OctopusNet addresses the problem of cross-modal interference, occurred in early-fusion, by extracting features from the modalities using separate modal encoders, which can be any CNN architecture, e.g. VGG [6], ResNet [7] or DenseNet [8]. As shown in Fig. 1(c), the feature maps generated at different stages of modal encoders are concatenated and fed to the hyper-fusion decoder. The proposed decoder uses hyper-fusion modules to fuse the feature maps from different modalities and avoid the problem of feature explosion. Compared to the late-fusion approach with four separate decoders, the hyper-fusion decoder more effectively fuses cross-modal information and reduces the computational cost. The decoder upsamples the high-level low-resolution feature maps back to the original resolution in the same way as [9], and yields the segmentation result.

2.1   Modal Encoder

As aforementioned, the modal encoder can be chosen from any widely-used network architectures, e.g. DenseNet. In our experiments, DenseNet-161 usually yields better segmentation accuracy compared to that of VGG and ResNet. Therefore, we take DenseNet-161 as an example to illustrate the pipeline of extracting feature maps from different modalities. The detailed information of network architecture of DenseNet-161 can be found in [8]. The colored cubes in each modal encoder in Fig. 1(c) are the feature maps generated by different stages of DenseNet-161, which correspond to the ones from Dense Block (1) - (4). To better leverage the explicit information contained in different modalities, all the extracted feature maps are concatenated together and fed to the hyper-fusion decoder for feature fusion and distilling.

2.2   Hyper-fusion Decoder

We propose a novel hyper-fusion decoder to decode and upsample the high-level low-resolution feature maps back to the original resolution of input and yield the segmentation result. As shown in Fig. 1(c), the decoder adopts a *hyper-fusion module* for feature distilling, *deep supervision* for better training convergence, and concatenates the fused feature maps (purple cubes) to the upsampled ones (yellow cubes) to produce segmentation result.

Hyper-fusion Module. As the network goes deeper, the number of feature maps increases. Consequently, the concatenation of multi-modal feature maps easily causes a problem of feature explosion. For example, assuming there are N modalities as input, the number of concatenated feature maps generated by Dense Block (4) of modal encoders is $2208 \times N$, which requires high cost of computation and memory consumption. The hyper-fusion module (1)–(4) in Fig. 1(c) is a 1×1 convolution, which has the same number of channels to that of feature maps from Dense Block (1)–(4). Therefore, the N concatenated feature maps can be accordingly fused and compacted to single ones using hyper-fusion modules (the purple cubes in Fig. 1(c)).

Deep Supervision. The proposed OctopusNet is an end-to-end framework, which means the multiple modal encoders and hyper-fusion decoder are simultaneously trained and updated. However, the networks adopted for modal encoders are usually extremely deep, resulting in a difficulty for training convergence only using a single supervision signal at the very end of a long pipeline. Hence, we added a weak supervision signal to the deepest node of OctopusNet, i.e. the elongated purple cube at the bottom. Assuming DenseNet-161 is adopted as the modal encoder, the size of bottom purple cube is $2208 \times 7 \times 7$. In this situation, a 1×1 convolution is used to transform the cube to $1 \times 7 \times 7$ and the original supervision signal is resized from 224×224 to 7×7 to be as weak supervision.

3   Experiments

We evaluate the performance of the proposed OctopusNet on publicly available datasets from two challenges, namely ISLES-2018[1] and MRBrainS-2013[2], and compare with the early-fusion and late-fusion approaches. Though our result is competitive to the top-performance of the ISLES-2018 challenge, the purpose of the experiments is not to win the challenges. Our main purpose is to demonstrate the effectiveness of the proposed fusion approach and compare with other widely used alternative strategies. Our approach is complementary to, and can be easily integrated into, other FCN based multi-modal segmentation approaches.

3.1   Datasets

ISLES-2018. Ischemic stroke lesions segmentation (ISLES) is a competition consecutively held since 2015 [10]. In ISLES-2018, the challenge organizer released a new dataset, which is composed of six modalities, including diffusion weighted imaging (DWI) MRI, computed tomography (CT) and four perfusion

---

[1] http://www.isles-challenge.org/.
[2] http://mrbrains13.isi.uu.nl/index.php.

scans, i.e. mean transit time (MTT), time to peak of the residue function (Tmax), cerebral blood flow (CBF) and cerebral blood volume (CBV). The ISLES-2018 competition provides 94 sets of multi-modal data for training and 62 sets for test. As the ground truth of the test set is not available, participants need to submit their prediction to the online system for performance evaluation.

The dataset has two main challenging issues. First, the appearances of lesion areas among different modalities are widely varied. The lesion area in MTT and Tmax is brighter than the normal area, while it is dark in the modalities of CBF and CBV, as shown in Fig. 1(c). Second, the ISLES-2018 data does not have a uniform size. Though each slice has a fixed size of 256×256 pixels, the number of slices contained in a volume varies from 2 to 22. Most ISLES-2018 volumes only have two slices, which presents a difficulty to adapt a 3D segmentation framework to the dataset.

MRBrainS-2013. The MRBrainS-2013 dataset contains five sets of multi-modal brain images, in which the brain tissues, i.e. gray matter, white matter and cerebrospinal fluid, are fully annotated. Three registered modalities, i.e. T1-weighted scan (T1), T1-weighted inversion recovery scan (T1_IR) and T2-weighted fluid attenuated inversion recovery scan (T2_FLAIR) are provided. The volumes of the dataset are in an uniform size of $240 \times 240 \times 48$ voxels.

Implementation Details. The proposed OctopusNet may have different architectures regarding to the input data. As most ISLES-2018 data has a couple of slices, we develop a 2.5D OctopusNet instead of 3D. Three consecutive slices from a volume are extracted and fed to 2D modal encoders as inputs. The first and last slices of the volume are duplicated for padding. In this setting, modal encoders can be pretrained on the ImageNet dataset for better training convergence. Our OctopusNet is implemented using PyTorch. The initial learning rate is set to 0.7 and divided by 10 after every 35 epochs. The network is optimized by stochastic gradient descent (SGD). The used datasets have different number of modalities. Consequently, the proposed OctopusNet involves different numbers of modal encoders for the ISLES-2018 and MRBrainS-2013.

Take the ISLES-2018 as an example. As the ISLES-2018 test set does not provide the DWI modality, five original modalities, i.e. CT, MTT, Tmax, CBF and CBV, are adopted as input for OctopusNet. Furthermore, the lesion area is not clearly visible in the modalities of CT, CBF and CBV. Hence, these three modalities are enhanced by histogram equalization. Finally, five original modalities and three enhanced modalities are fed to the OctopusNet[3]. The input size of each modal encoder is $256 \times 256 \times 3$.

3.2 Performance Analysis

We perform a five-fold cross validation on the ISLES-2018 and MRBrainS-2013 training set to evaluate the performance of our OctopusNet. All the experiments

---

[3] This network has an octopus shape with a body (the decoder) and eight arms (the encoders). This is where the name, OctopusNet, comes from.

Table 1. Dice coefficient (%) of lesion areas of ISLES-2018 (average of five-fold cross validation).

	VGG-16 [6]	ResNet-50 [7]	DenseNet-161 [8]
Single modality (Tmax)	44.97	44.03	45.83
Early-fusion	53.38	53.99	53.82
Late-fusion	53.73	55.39	53.86
Octopus-fusion	**55.71**	**57.33**	**57.72**
Octopus-fusion + deep supervision	-	-	57.90

are repeated three times to reduce the influence caused by random nature of network training. Hence, the results reported in the paper are the the average results of three repeated experiments. For the convenience of comparison, the frameworks using baseline fusion approaches (early- and late-fusion) in our experiments are in the same setting to that of OctopusNet, e.g. the input format of different modalities. Henceforth, the fusion approach adopted in our Octopus-Net is named as Octopus-fusion. The Dice coefficient, which measures the spatial overlap index between the segmentation results and ground truths, is adopted as the metric to evaluate the segmentation accuracy.

Results on ISLES-2018. As aforementioned, the modal encoder can be chosen from widely used deep learning networks. To evaluate the generalization capability of Octopus-fusion, several network architectures, e.g. VGG-16 [6], ResNet-50 [7] and Dense-Net-161 [8], are adopted as the modal encoder and trained with different fusion approaches on the ISLES-2018 dataset. The results are listed in Table 1. To evaluate the improvement produced by the usage of multi-modal images, we also report segmentation accuracy using a single modality. Due to the space limit, Table 1 only lists the result of the best single modality (i.e., Tmax). Due to the lack of information contained in extra modalities, the frameworks using single modality only yield Dice coefficients around 44%, which are about 9% lower than that of multi-modal frameworks. For the early-fusion approach, Table 1 shows that the accuracies of all three backbone networks are quite similar with the deep networks (i.e., DenseNet-161 and ResNet-50) slightly outperforming the shallow network of VGG-16.

For the late-fusion approach, as it involves multiple encoder-decoder architectures for different modalities, the explicit information contained in multi-modal data can be better extracted. Hence, accuracy of the late-fusion approach surpasses that of early-fusion with the same modal encoders. However, DenseNet-161 only gains marginal improvement, i.e. 0.04%, by switching from early-fusion to late-fusion. The reason for that is the network depth of DenseNet-161 is extremely deep, which makes it difficult to simultaneously well train multiple fully convolutional DenseNet-161 branches in the late-fusion approach. Compared to the results of early- and late-fusion, our Octopus-fusion approach significantly boosts the accuracy of modal encoders. The best lesion segmentation result is achieved by

Table 2. Dice coefficients (%) yielded by different fusion approaches for each brain tissue of MRBrainS-2013 (average of five-fold cross validation).

	CSF	Gray matter	White matter	Ave. Dice
Single modality (T1)	78.46	81.37	85.69	81.84
Single modality (T1_IR)	75.95	77.51	81.92	78.46
Single modality (T2_FLAIR)	74.00	75.63	77.32	75.65
Early-fusion	79.05	80.58	83.56	81.07
Late-fusion	79.16	81.41	84.71	81.76
Octopus-fusion	**80.59**	**82.12**	**86.05**	**82.92**

the Octopus-fusion DenseNet-161, i.e. a Dice of 57.72%. By adding the *deep supervision* signal, the segmentation accuracy is further increased to 57.90%, which is 2.51% higher than that of the best-performance among benchmarking algorithms (late-fusion with ResNet-50).

ISLES-2018 Challenge. We participated the ISLES-2018 competition. The proposed OctopusNet using DenseNet-161 achieved an average Dice of 48%, which ranked the third-place of ISLES-2018 challenge[4]. We notice that, for all participating teams, there is a gap between validation and test accuracy. One possible reason is that the test set contains more small lesions, where are difficult to segment accurately for all algorithms. Additionally, the top approaches reported that they used extra modalities, e.g. 4D perfusion CT (ranked 1st with Dice of 51%) and synthesized DWI (ranked 2nd with Dice of 49%), which were not adopted by our OctopusNet.

Results on MRBrainS-2013. We also conduct experiments on MRBrainS-2013 to compare the performances of different fusion approaches for the task of brain tissue segmentation. The three original modalities of MRBrainS-2013 are directly employed as input for the proposed OctopusNet. The best-performer on ISLES-2018, i.e. DenseNet-161, is adopted as the backbone of modal encoder. The input size of each modal encoder is 240 × 240 × 3. The Dice coefficients for different tissues, including CSF, gray matter and white matter, produced by different fusion approaches are listed in Table 2. The average Dice (Ave. Dice) is calculated by averaging the Dice coefficients of three tissues.

The framework using single modality is also evaluated for comparison. It is interesting to see that, for the gray matter and white matter, the best single modality (T1) produces even higher segmentation accuracy than early-fusion. The reason for that may be the physicians mark the annotation of the gray and white matter primarily using the T1 scans, while the T1_IR and T2_FLAIR scans usually provide additional information for the annotation of outer border of CSF and white matter lesion, respectively. Therefore, most information contained in the extra modalities, i.e. T1_IR and T2_FLAIR, may be seen as noises for the

---

[4] https://www.smir.ch/ISLES/Start2018.

brain tissue segmentation. The late-fusion approach yields similar average segmentation accuracy to using T1 only (81.76% vs. 81.84%). The reason for that may be the post-fusion approach performs information fusion too late; Therefore, it can not fully utilize the complementary information among multiple modalities. Oppositely, by using the proposed Octopus-fusion, the average segmentation accuracy increases to 82.92% and improvement is observed for all tissues, which illustrates that our Octopus-fusion can effectively extract useful information from each modality and prevent the cross-modal interference caused by irrelevant information. An additional observation is that CSF consistently benefits from multi-modality fusion using any fusion strategy, which is concordant to the annotation process of physicians. Again, Octopus-fusion achieves the largest boost in segmentation accuracy of CSF, i.e., +2.13%.

4   Conclusion

In this paper, we presented a novel deep learning network architecture, namely OctopusNet, for multi-modal medical image segmentation. The proposed OctopusNet adopted a separate modal encoder for each modality to explicitly extract features and a hyper-fusion decoder to fuse the features, avoiding the problem of feature explosion. The proposed OctopusNet was evaluated on two publicly available datasets. The experimental results demonstrated that our OctopusNet was a general network architecture, which can provide excellent performance for various segmentation tasks of multi-modal medical data.

References

1. Pereira, S., Alves, V., Silva, C.A.: Adaptive feature recombination and recalibration for semantic segmentation: application to brain tumor segmentation in MRI. In: Frangi, A.F., Schnabel, J.A., Davatzikos, C., Alberola-López, C., Fichtinger, G. (eds.) MICCAI 2018. LNCS, vol. 11072, pp. 706–714. Springer, Cham (2018). https://doi.org/10.1007/978-3-030-00931-1_81
2. Shen, H., Wang, R., Zhang, J., McKenna, S.J.: Boundary-aware fully convolutional network for brain tumor segmentation. In: Descoteaux, M., Maier-Hein, L., Franz, A., Jannin, P., Collins, D.L., Duchesne, S. (eds.) MICCAI 2017. LNCS, vol. 10434, pp. 433–441. Springer, Cham (2017). https://doi.org/10.1007/978-3-319-66185-8_49
3. Nie, D., Wang, L., Gao, Y., Shen, D.: Fully convolutional networks for multi-modality isointense infant brain image segmentation. In: ISBI, pp. 1342–1345 (2016)
4. Wang, L., et al.: Volume-based analysis of 6-month-old infant brain MRI for autism biomarker identification and early diagnosis. In: Frangi, A.F., Schnabel, J.A., Davatzikos, C., Alberola-López, C., Fichtinger, G. (eds.) MICCAI 2018. LNCS, vol. 11072, pp. 411–419. Springer, Cham (2018). https://doi.org/10.1007/978-3-030-00931-1_47
5. Wu, Z., et al.: Registration-free infant cortical surface parcellation using deep convolutional neural networks. In: Frangi, A.F., Schnabel, J.A., Davatzikos, C., Alberola-López, C., Fichtinger, G. (eds.) MICCAI 2018. LNCS, vol. 11072, pp. 672–680. Springer, Cham (2018). https://doi.org/10.1007/978-3-030-00931-1_77

6. Simonyan, K., Zisserman, A.: Very deep convolutional networks for large-scale image recognition. arXiv e-print arXiv:1409.1556 (2014)
7. He, K., Zhang, X., Ren, S., Sun, J.: Deep residual learning for image recognition. In: CVPR, pp. 770–778 (2016)
8. Huang, G., Liu, Z., Maaten, L.V.D., Weinberger, K.Q.: Densely connected convolutional networks. In: CVPR, pp. 2261–2269 (2017)
9. Ronneberger, O., Fischer, P., Brox, T.: U-net: convolutional networks for biomedical image segmentation. In: Navab, N., Hornegger, J., Wells, W.M., Frangi, A.F. (eds.) MICCAI 2015. LNCS, vol. 9351, pp. 234–241. Springer, Cham (2015). https://doi.org/10.1007/978-3-319-24574-4_28
10. Maier, O., Menze, B.H., Gablentz, J.V.D., et al.: ISLES 2015 - a public evaluation benchmark for ischemic stroke lesion segmentation from multispectral MRI. Med. Image Anal. **35**, 250–269 (2017)

Neural Architecture Search for Optimizing Deep Belief Network Models of fMRI Data

Ning Qiang[1], Bao Ge[1($\boxtimes$)], Qinglin Dong[2], Fangfei Ge[2],
and Tianming Liu[2]

[1] Shaanxi Normal University, Xi'an, China
Bob_ge@snnu.edu.cn
[2] The University of Georgia, Athens, GA, USA

Abstract. It has been shown that deep neural networks are powerful and flexible models that can be applied on fMRI data with superb representation ability over traditional methods. However, a new challenge of neural network architecture design has also attracted attention: due to the high dimension of fMRI volume images, the manual process of network model design is very time-consuming and error prone. To tackle this problem, we proposed a Particle Swarm Optimization (PSO) based neural architecture search (NAS) framework for a deep belief network (DBN) that models volumetric fMRI data, named NAS-DBN. The core idea is that the particle swarm in our NAS framework can temporally evolve and finally converge to a feasible optimal solution. Experimental results showed that the proposed NAS-DBN framework can find robust architecture with minimal testing loss. Furthermore, we compared functional brain networks derived by NAS-DBN with general linear model (GLM), and the results demonstrated that the NAS-DBN is effective in modeling volumetric fMRI data.

Keywords: Neural Architecture Search (NAS) · Particle swarm optimization (PSO) · Deep Belief Network · Task fMRI

1 Introduction

Understanding the organizational architecture of functional brain networks has raised intense interest since the inception of neuroscience [1]. In recent years, deep learning has attracted much attention in the field of machine learning and data mining, and it has been demonstrated to be a powerful tool for modeling brain networks based on fMRI data, compared to traditional shallow methods such as general linear model (GLM) [2], and independent component analysis (ICA) [3], and sparse dictionary learning (SDL) [4]. Although deep learning has enjoyed remarkable progresses over the past few years, most current neural network architectures were developed manually by researchers, which typically is a very time-consuming and error prone process, since all hyper-parameters of neural networks were decided by expert experiences. Fortunately, Neural Architecture Search (NAS), aiming to automatically search for optimal network architecture, is recently considered as a feasible and promising solution to the abovementioned problem. During recent years, several novel NAS methods, e.g., either

© Springer Nature Switzerland AG 2020
Q. Li et al. (Eds.): MMMI 2019, LNCS 11977, pp. 26–34, 2020.
https://doi.org/10.1007/978-3-030-37969-8_4

based on reinforcement learning or evolutionary computation, have been developed and applied in a variety of deep learning tasks [5]. However, due to the high dimension and complexity of volumetric fMRI data, there is still few NAS applications in the field of brain imaging using fMRI.

To fill the above gap, in this work, we firstly propose a novel multi-layer volumetric deep belief network (DBN) and designed a group-wise scheme that aggregated multiple subjects' fMRI volume data for effective training of the DBN, with the purpose of discovering meaningful functional brain networks (FBN) in task-based fMRI data. Secondly, and more importantly, aiming to find out the optimal network architecture of DBN in modeling fMRI volumes, we developed a novel NAS framework based on particle swarm optimization (PSO). The key idea is that the particle swarm in the NAS framework will temporally evolve and finally converge to a feasible optimal solution. To quantitatively evaluate the performance of the NAS-DBN framework, a series of experiments have been conducted and the results showed the effectiveness of our design. Furthermore, we used the DBN with optimal architecture to extract FBNs from task-based fMRI data of Human Connectome Project (HCP) and compared the results with GLM-derived brain networks. Our results demonstrated that the NAS-DBN is a promising tool for deriving meaningful and interpretable FBNs from fMRI data.

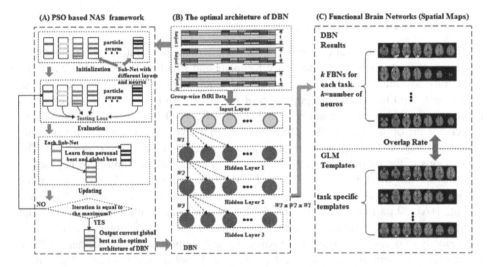

Fig. 1. Illustration of proposed NAS-DBN framework for deriving functional brain networks from task-based fMRI data

2 Materials and Methods

2.1 Overview

Figure 1 summarizes our PSO-based NAS framework (Fig. 1(A)) and DBN structure (Fig. 1(B)) for modeling FBNs. The particle swarm consists of 30 particles, each of which represents a subnet with different initial architecture (Fig. 1(A)). We investigated two main hyper-parameters including the number of layer and the number of neurons in each layer. These two parameters are used to construct a mapping between a particle position and a solution of network architecture design. The testing loss of DBN is regarded as the fitness function of PSO, which will be minimized in the searching process. The particle swarm can evolve and converge to an optimal solution. Then we applied this optimal architecture of DBN to model FBNs from task-based fMRI data (Fig. 1(B)), and the weights of network are visualized and quantified as FBNs (Fig. 1 (C)), which will be further compared with GLM-derived network maps.

2.2   Dataset and Preprocessing

In this paper, fMRI data from the Human Connectome Project (HCP) 900 Subjects Release was adopted as training dataset. The stimuli were projected onto a computer screen behind the subject's head within the imaging chamber, and 4 out of 7 categories of behavioral tasks are used, including Emotion, Gambling, Language, and Social. The fMRI preprocessing pipelines were implemented by FSL FEAT (FMRIB's Expert Analysis Tool) and Nilearn [6], including spatial resampling to the MNI152 template, frequency filtering, detrending, normalization and masking. The details of acquisition parameters and information of each task can be found in the literature [4].

2.3   PSO Based NAS Framework

Particle Swarm Optimization (PSO) is a swarm based evolutionary computation algorithm that is originally proposed by Kennedy and Eberhart in 1995 [7]. Due to its numerous advantages, such as less parameter requirements, simple formula, easy to implement, PSO has become a popular tool for solving various complex optimization problems. In this work, we adopted and designed a PSO based NAS framework to search for the optimal network architecture of DBN. We designed a two-dimensional encoding method to map network architecture of DBN to a particle position. The dimensions of the particle represent the number of layer and the number of neurons in each layer with the range of (2, 10) and (20, 200), respectively. In order to reduce computational cost, we assume the number of neurons in each layer is equal. As shown in Fig. 1(A), 30 particles are initialized in the solution space with initial velocities and positions. A particle position represents a solution of network architecture design, and the velocity of particle determines the particle's next motion, which is affected by three factors: current motion, personal best position and global best position. The whole swarm is attracted by the global best and is exploring in the solution space, and at the same time each particle is exploiting its nearby space because of attraction of personal best. The process of exploring and exploiting also has a randomness, making PSO a stochastic and intellectual searching algorithm, thus the whole swarm can quickly converge to a feasible optimal solution compared to other exhaustive search algorithms.

The evolutionary process of particle swarm mainly consists of two steps: evaluation and updating. First, after initialization, all particles are evaluated by a fitness function which is defined by the testing loss of DBN. To avoid potential overfitting in NAS process, testing loss is adopted instead of training loss as an evaluation index of the model. After training, the trained model is applied to predict testing data (not used in training) and the Mean Squared Error (MSE) between input and output is calculated as testing loss, also the fitness value of corresponding particle. The split ratio of training and testing set was set as 0.2. Notably, the input data was normalized to a Gaussian distribution for effective training. Then the personal best solution of each particle and the global best solution of whole swarm are recorded. Second, all particles' velocities and positions are updated by the following equations:

$$v_{id}^{t+1} = w \cdot v_{id}^{t} + c_1 \cdot r_1 \left(p_{id}^{t} - x_{id}^{t} \right) + c_2 \cdot r_2 \left(p_{gd}^{t} - x_{id}^{t} \right) \tag{1}$$

$$x_{id}^{t+1} = x_{id}^{t} + v_{id}^{t+1} \tag{2}$$

Equations (1) and (2) are for velocity and position updating, respectively, where x_{id}^{t} and x_{id}^{t+1} are the current and next positions, respectively; v_{id}^{t} and v_{id}^{t+1} are the current and next velocities, respectively. The subscripts t, i, and d denote current iteration, subnet, and coding dimension, respectively; w is the inertia weight that reflects the inertia of particle motion; $c1$ and $c2$ are learning rate that affect the ratio of learning towards personal best and global best, making the searching process intelligent; $r1$ and $r2$ are two uniform random numbers selected from the interval [0, 1], which give the searching process a certain randomness. The second and third parts of the right side of Eq. (1) reflect that current particle's next motion is affected by personal best position (p_{id}^{t}) and global best position (p_{gd}^{t}), as well as its previous motion. In addition, a uniform mutation strategy with variable mutation probability was introduced to increase the diversity of particle swarm. At the beginning of iteration, greater mutation probability makes the algorithm to have better exploration ability, and smaller mutation probability makes the algorithm to have better exploiting ability in the last stage of iteration. Therefore, the mutation probability was set to a linearly changing value from 0.2 to 0.05 with the increase of iteration number. Notably, we perform convergence check after initialization and updating, since some of particles might be divergent in the training process. These non-convergent subnets will be replaced by re-initiated subnets, and they will also be checked until they converge.

2.4   DBN Model of Volumetric FMRI Data

DBN, constructed by blocks of Restricted Boltzmann Machines (RBM), is widely used for deep generative models and has been proven to be a powerful tool for modeling fMRI data. Here, a group-wise volumetric scheme of DBN is proposed to model fMRI volumes. Considering the large inter-subject variability among human brains, arbitrary selection of a single individual may not effectively represent the population, thus a group-wise learning scheme is needed to reduce inter-subject variability by jointly registering the fMRI volumes to a common reference template corresponding to the group average. Since that the inter-subject variability is relatively more associated with the volatile time courses in different imaging sessions, it appears that taking volumes as input possibly works better than time series in terms of modeling the FBNs from fMRI data in this work. Accordingly, a volume from the fMRI data was taken as a feature, each time frame was taken as a sample, and a group-wise temporal concatenation was applied to all HCP subjects.

In particular, to reduce the possibility of overfitting and to improve generalization, a sparse weight regularization was designed and added in the DBN model. In each iteration, the weight is updated with the estimated gradient and an extra term of weight regularization derivative. In this paper, L1 weight penalty served as the regulation term while calculating the derivative of the sum of the absolute values of the weights. The sparse weight regularization works by causing many of the weights to become zero while allowing a few of the weights to grow large. In the context of fMRI data, L1 regularization can denoise the FBNs and improve interpretability by suppressing useless weights and allowing important model parameters to become larger, which is considered as an important methodology/technical contribution. In this work, L1 regularization was empirically set as 0.00001.

With respect to interpreting a trained DBN in the fMRI context, each row of weight vector was mapped back into the original 3D brain image space, which was the inverse operation of masking in preprocessing steps and was interpreted as an FBN. After the DBN is trained layer-wisely on a large-scale task fMRI dataset, each weight showed the extent of each voxel contributed to a latent variable. For deeper layers, the linear combination approach was used to interpret the connection. With this approach, as an example, $W_3 \times W_2 \times W_1$ was visualized for the first hidden layer as FBNs (Fig. 1(C)).

2.5   Implementations

The NAS-DBN is inherently much more computationally expensive, compared to DBN models for temporal fMRI time series. Considering HCP 4D images and one single layer of RBM, there are around 20K trainable parameters for temporal fMRI time series DBN, but 20 million for volumetric fMRI DBN. Moreover, the population size and iteration size will put significant computational burden on the NAS process. To deal with this problem, in this paper, the TensorFlow, which is a popular deep learning framework and provides great convenience coding with GPUs, was adopted with high efficiency GPU computation to fill the gap. Based on TensorFlow, we designed and implemented a fast and flexible DBN model. Limited by computing resources, all subnets will be trained one by one and processed collectively. The code was run on a deep learning server with GeForce GTX 1080 TI of GPU and 32 Gb of RAM.

3   Results

3.1   Comparisons Between NAS-DBN and DBN

To quantitatively evaluate the effectiveness of our NAS-DBN framework, we ran 10 times of the searching process independently, and analyzed the statistical results. We used 4 shuffled HCP tasks data as input of NAS. After NAS, we used the same optimal architecture of DBN to model each task data independently. As shown in Fig. 2, the optimal results show high consistence and robustness in the optimal number of layer and the optimal number of neurons. In most runs (8 out of 10), the result of the optimal number of layer is 3, except only two results are 2 and 4 respectively. The optimal number of neurons is 80, and all results are in a range from 71 to 112. These statistical experiments demonstrated that our NAS framework can generate reliable results of architecture design. Furthermore, we compared the testing loss of DBNs with optimal architecture and manually selected architectures. Figure 3 shows comparison of testing loss of 4 DBNs with the same number of neurons. Figure 4 shows comparison of testing loss of 4 DBNs with the same layers. DBN (3,80) denotes that there are 3 hidden layers and 80 neurons in this DBN structure. It can be seen that DBN with the optimal architecture from NAS has the lowest testing loss of 0.0213 compared to other manually designed DBNs, demonstrating the effectiveness of NAS framework.

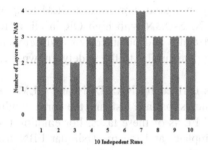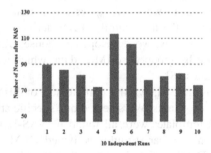

Fig. 2. Statistical results of 10 independent experiments in the optimal number of layers and optimal number of neurons.

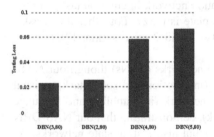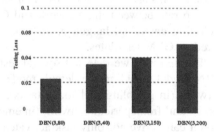

Fig. 3. Testing loss of DBNs with same neurons.

Fig. 4. Testing loss of DBNs with the same layers

3.2   Comparison of NAS-DBN with GLM

To explore the representation of task-based fMRI data, four task-specific DBNs were trained on fMRI data of 4 HCP tasks independently using the same hyperparameters. To quantitatively evaluate the performance of DBN in modeling tfMRI data, a comparison study between NAS-DBN results and the widely known GLM activation results is investigated in this section. For fare comparison, all the functional networks derived by these two methods are thresholded at Z > 2.3 after transformation into "Z-scores" across spatial volumes. The spatial overlap rate is defined to measure the similarity of two FBNs in accordance with previous literature studies. Here, the spatial similarity is defined by the overlap rate (OR) between two functional networks $N^{(1)}$ and $N^{(2)}$ as follows, where n is the volume size:

$$OR\left(N^{(1)}, N^{(2)}\right) = \frac{\sum_{i=1}^{n} \left| N_i^{(1)} \cap N_i^{(2)} \right|}{\sum_{i=1}^{n} \left| N_i^{(1)} \cup N_i^{(2)} \right|} \tag{3}$$

With the similarity measure defined above, the similarities $OR(N_{DBN}, N_{GLM})$ between the NAS-DBN derived functional networks N_{DBN} and the GLM derived functional networks N_{GLM} were quantitatively measured. For each of GLM template, we found the most similar FBN derived by NAS-DBN with high OR in all 4 HCP tasks. Notably, we developed in-house GLM codes and obtained our own templates derived by group-wise fMRI data, which are quite similar to the widely known GLM templates [8].

Figure 5 shows the comparison of FBNs derived by NAS-DBN and GLM templates in 4 tasks. We selected one specific stimulus for each task and the corresponding GLM templates were all found in NAS-DBN FBNs of these tasks. For emotion task, we can see fear stimulus activated GLM template, and the most similar FBN from NAS-DBN, which is the 12th network out of 80 networks. Comparing this brain network with the benchmark GLM template, the overlap rate is as high as 0.502, and thus it is easy to recognize the close match between them. Since our NAS-DBN is an unsupervised architecture, there are other similar FBNs that can be detected in all 4 tasks. For instance, we found a similar fear activation network in emotion task, and the overlap rate between this network and GLM template is 0.327. For other three tasks, including gambling, language, and social, we also found 2 most similar FBNs compared to GLM templates.

Furthermore, we detected several resting state networks (RSNs) though our NAS-DBN model including the default mode network (in emotion and social tasks), visual network (in gambling and social tasks), auditory network (in gambling and language tasks), and frontoparietal network (in emotion task), demonstrating that our NAS-DBN model can derive not only task activated networks but also resting state networks. As shown in Fig. 6, 4 RSNs were found and visualized in our DBN-derived FBNs in different tasks. Here, we used the RSN templates from Nilearn [6] as benchmark.

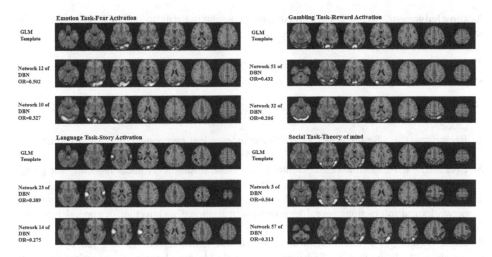

Fig. 5. Comparison between GLM templates and similar FBNs derived by NAS-DBN in emotion, gambling, language, and social tasks. Each network is visualized with 7 axial slices.

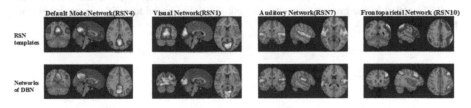

Fig. 6. Comparison between RSN templates and similar FBNs derived by NAS-DBN in different tasks. Each network is visualized with 3 most informative orthogonal slices.

4 Conclusion

We proposed a novel PSO based NAS-DBN framework for searching optimal architecture of DBN in modeling FBNs from volumetric fMRI data. Four HCP fMRI datasets were used to validate our NAS-DBN. Based on evolutionary computation, 30 subnets in our framework learn experience of each other, and the whole swarm can evolve and finally converge to a feasible optimal architecture of DBN. We selected testing loss as fitness function of NAS, instead of training loss, to avoid potential overfitting in NAS process. The statistical experiment of NAS showed high consistence and robustness of our architecture design. Furthermore, a comparison between GLM and DBN validated that the functional networks learned by NAS-DBN are meaningful and can be well interpreted. We also showed that our DBN model is capable of deriving both task specific functional networks and resting state networks. The promising results by NAS-DBN model showed the importance of optimizing neural network structures in deep learning.

Acknowledgement. We thank all investigators contributing data to the HCP project. Bao Ge was supported by NSFC61976131.

References

1. Logothetis, N.K.: What we can do and what we cannot do with fMRI. Nature **453**(7197), 869 (2008)
2. Beckmann, C.F., Smith, S.M.: Probabilistic independent component analysis for functional magnetic resonance imaging. IEEE Trans. Med. Imaging **23**(2), 137–152 (2004)
3. Beckmann, C.F., Jenkinson, M., Smith, S.M.: General multilevel linear modeling for group analysis in FMRI. Neuroimage **20**(2), 1052–1063 (2003)
4. Lv, J., Jiang, X., Li, X., Zhu, D., Zhang, S., Zhao, S., et al.: Holistic atlases of functional networks and interactions reveal reciprocal organizational architecture of cortical function. IEEE Trans. Biomed. Eng. **62**(4), 1120–1131 (2015)
5. Zoph, B., Le, Q.V.: Neural architecture search with reinforcement learning (2016)
6. Abraham, A., Pedregosa, F., Eickenberg, M., et al.: Machine learning for neuroimaging with scikit-learn. Front. Neuroinformatics **8**, 14 (2014)
7. Kennedy, J.: Particle swarm optimization. Encycl. Mach. Learn. **4**, 760–766 (2010)
8. Barch, D.M., Burgess, G.C., Harms, M.P., Petersen, S.E., Schlaggar, B.L., Corbetta, M., et al.: Function in the human connectome: task-fMRI and individual differences in behavior. Neuroimage **80**, 169–189 (2013)

Feature Pyramid Based Attention
for Cervical Image Classification

Hongfeng Li[1,3], Jian Zhao[2], Li Zhang[1,3], Jie Zhao[1], Li Yang[4(✉)],
and Quanzheng Li[5(✉)]

[1] Center for Data Science in Health and Medicine, Peking University, Beijing, China
[2] Department of Gynaecology and Obstetrics, Peking University First Hospital,
Beijing, China
[3] Beijing Institute of Big Data Research, Beijing, China
[4] Department of Nephrology, Peking University First Hospital, Beijing, China
li.yang@bjmu.edu.cn
[5] Massachusetts General Hospital and Harvard Medical School,
Gordon Center for Medical Imaging, Boston, MA 02114, USA
quanzhengli5@gmail.com

Abstract. It is known that cervical cancer has been a great threaten
to the health of women worldwide. Regular examination through images
obtained by colposcopes can facilitate early detection and treatment.
However, it is challenging for computer-aided methods to perform diagnosis with the cervical images correctly. To this end, we develop a novel
deep neural network based method for cervical image classification in this
paper. The proposed method first generates a sequence of feature maps
through lateral connection and feature fusion so as to fully exploit the
multi-scale information included in a main network. Then compatibility
scores are computed between the multiple feature maps and the global
feature outputted by the main network, with which attention maps can
be obtained. The attention maps are able to extract salient features from
images via focusing on the crucial regions of the cervixes, which can be a
guidance for clinical diagnosis. We test the proposed method on CIFAR
datasets and a cervical dataset collected from the Peking University First
Hospital and results show that the proposed method achieves excellent
performances and outperforms the related approaches.

Keywords: Cervical image classification · Deep neural networks ·
Attention

1 Introduction

Cervical cancer is one of the most common gynecological malignances, and its
incidence is only lower than that of breast cancer in female reproductive system
tumors [2]. Therefore, it is necessary to conduct a gynecological examination

---

H. Li and J. Zhao are equally contributed.

© Springer Nature Switzerland AG 2020
Q. Li et al. (Eds.): MMMI 2019, LNCS 11977, pp. 35–42, 2020.
https://doi.org/10.1007/978-3-030-37969-8_5

with the help of colposcopy equipments such that early detection, diagnosis and treatment of cervical cancer can be achieved. However, due to the inferior quality of images obtained and variance in cervical morphology, it is challenging to diagnose cervical cancer correctly with images obtained by colposcopy equipments.

Recently, developing effective algorithms for medical image processing has become one of the most important research directions in the field of machine learning. Utilizing machine learning methods to analyze and process cervical images can provide a valuable reference for clinicians to diagnose and prevent diseases. However, classifying colposcopy images with machine learning algorithms remains a challenging problem.

Existing classification of colposcopy images is heavily dependent on the experience of doctors. Examining large numbers of colposcopy images manually is tedious and time-consuming, and the examination results depend heavily on individual experience. By contrast, automatic classification methods can bring more stable results, which can be a huge clinical demand. There have been a few works for cervical image classification currently. However, these works mainly use traditional machine learning methods and lots of human prior information is required, which is difficult to generalize [6]. Recently, deep learning has been widely applied in the area of medical image analysis, such as image classification [9], segmentation [5] and image restoration [1]. Particularly, a few deep learning methods have also been used for colposcopy images processing [10,13]. Nevertheless, these methods are often just an application of existing algorithms to the related problems and innovation on theory or methodology is lacked.

Attention mechanisms were first widely applied to natural language processing [12,14]. The mechanism of attention is to learn a context vector to weight the input so as to highlight the salient features while suppress the unrelated counterparts. In this way, the prediction can be more targeted. Recently, there are a few attention based algorithms proposed for medical image classification [15]. However, these methods often need bounding box labels or the context information is obtained through region proposal and hard-cropping [4].

To resolve the above issues, we propose a feature pyramid based attention method for cervical image classification in this paper. The method can learn salient feature of images through fusing information at different scales of a deep network in an end-to-end way and do not require any bounding box labels or region proposal. Experiments on CIFAR datasets and a real cervical image dataset show that the proposed method can achieve excellent performances.

To summarize, the highlights of this paper can be listed as follows.

(1) A novel deep neural network with attention mechanism is proposed for cervical image classification.
(2) To fully utilize the multi-scale information included in the deep network, we construct a feature pyramid through lateral connections and feature fusion. With the obtained feature maps, attention maps are learned at different scales to extract salient features. Particularly, the attention construction methodology can be applied to any other deep networks.

(3) The proposed method achieves excellent performances on a real cervical image dataset, and thus is promising for practical cervical image classification tasks.

The remainder of the paper is organized as follows. Section 2 details the proposed method; Then we test the proposed method on CIFAR datasets and a real cervical image dataset and present the experimental results in Sect. 3; Finally, a conclusion is drawn in Sect. 4.

2 The Proposed Method

In this section, we detail the construction of the proposed method. The framework of the proposed method is illustrated in Fig. 1.

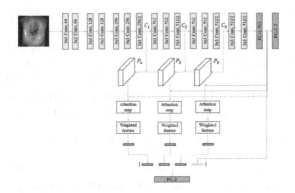

Fig. 1. The framework of the proposed method.

2.1 Feature Pyramid Construction

Similar to [8], we first construct a feature pyramid to exploit the information contained at different layers of a network. The VGG-16 network [11] is employed as the main network here. Particularly, each of the first two max-pooling layers of the original architecture is moved after each of the two corresponding additional convolutional layers introduced at the end of the pipeline as [4].

As Fig. 1 shows, we input a single-scale image to the main network and obtain a sequence of feature maps with various sizes. There are often many different layers outputting feature maps with the same sizes and we say these layers are in the same stage of the network. To construct the feature pyramid, only the last feature map in each stage is chosen. Specifically, the outputs of the 7-th, 10-th and 13-th layers of the VGG-16 network are selected to construct the feature pyramid and denoted as C_1, C_2 and C_3 respectively. Note that the spatial size of C_1 is the same as that of the input image and the spatial sizes of C_2 and C_3 are $\frac{1}{2}$ and $\frac{1}{4}$ of that of the input image respectively.

As Fig. 1 shows, the coarser feature maps at the higher layers are upsampled with a factor of 2 (bilinear interpolation for simplicity). Then they are merged with the corresponding feature maps with the same sizes from the main network by element-wise addition. The process is iterated until the feature map with the finest resolution is applied. Specifically, we start the iteration by simply taking the feature map C_3 as the coarsest resolution map P_3. Then, P_3 is upsampled and merged with the feature map C_2. The resulting feature map is denoted as P_2. The finest merged feature map P_1 is obtained in the similar way. The final feature maps P_1, P_2 and P_3 are of the same spatial sizes as those of the feature maps C_1, C_2 and C_3 respectively. It should be noted that, different from the feature construction process in [8], feature map smoothing operation is not applied to P_1, P_2 and P_3 since we do not observe any benefit in experiments.

2.2   Attention Modules Construction

We denote the feature vectors in each merged feature map $P_i, i = 1, 2, 3$ by $\mathcal{V} = \{\mathbf{v}_i^1, \mathbf{v}_i^2, \cdots, \mathbf{v}_i^n\}$. Here, $\mathbf{v}_i^j$ is the vector located at the i-th spatial position of P_i and n is the total number of vectors. We define the global feature vector $\mathbf{g}$ as the output of the main network which is just before the last fully-connected layer producing the prediction score. Assume that there is a compatibility function $\mathcal{F}$ which tasks two vectors with equal dimension as inputs and outputs a scalar compatibility score. We utilize the following function in this paper:

$$f_i^j = \langle \mathbf{w}, \mathbf{v}_i^j + \mathbf{g} \rangle, i = 1, 2, 3 \text{ and } j = 1, 2, \cdots, n \tag{1}$$

where $\mathbf{w}$ is the weight that can be learned and f_i^j is the compatibility score corresponding to $\mathbf{v}_i^j$. The weight can be interpreted as learning a set of features that are most related to the objects in an image.

For each merged feature map $P_i, i = 1, 2, 3$, the corresponding set of compatibility scores are computed as $\mathcal{F}(\hat{P}_i, \mathbf{g}) = \{f_i^1, f_i^2, \cdots, f_i^n\}$, where $\hat{P}_i$ is obtained by performing a linear transform to P_i. The transform maps the vector $\mathbf{v}_i^j$ to the dimension of $\mathbf{g}$. Then, the compatibility scores are normalized as follows:

$$a_i^j = \frac{e^{f_i^j}}{\sum_{j=1}^n e^{f_i^j}}, i = 1, 2, 3 \text{ and } j = 1, 2, \cdots, n \tag{2}$$

The normalized compatibility scores $\mathcal{A}_i = \{a_i^1, a_i^2, \cdots, a_i^n\}$ are utilized to compute a vector representing the semantical features of feature map P_i:

$$\mathbf{g}_i = \sum_{j=1}^n a_i^j \mathbf{v}_i^j, i = 1, 2, 3 \text{ and } j = 1, 2, \cdots, n \tag{3}$$

$\mathbf{g}_i$ can be taken as a descriptor of the input and $\mathcal{A}$ is interpreted as "attention".

Then the descriptors $\mathbf{g}_1$, $\mathbf{g}_2$ and $\mathbf{g}_3$ together with the global descriptor $\mathbf{g}$ are concatenated to form the final descriptor of the input image as

$$\mathbf{g}_a = [\mathbf{g}_1, \mathbf{g}_2, \mathbf{g}_3, \mathbf{g}] \tag{4}$$

The final image descriptor $\mathbf{g}_a$ is normalized and input to a fully-connected layer to obtain the class predictions for the input image.

As [4] shows that it is beneficial to identify the salient regions in an image and amplify the corresponding influence when making decisions. Compared with the method in [4], the proposed method is able to make full use of the multi-scale intermediate feature maps to extract more salient features from images. Moreover, it is beneficial to include the global feature $\mathbf{g}$ for final prediction since it contains meaningful high-level information of the input image.

3 Experiments and Analysis

In this section, we evaluate the performance of the proposed method on three datasets: the CIFAR-10 dataset, CIFAR-100 dataset [7] and a cervical image dataset collected from the Peking University First Hospital. In addition, we compare it with other related methods, including standard VGG-16 network [11] and the attention method in [4]. The experiments are performed on a server with a GTX 1080Ti GPU. The proposed algorithm is trained with 300 epochs in total and a mini-batch size of 12. The initial learning rate is 0.1 and dropped by 0.1 times every 25 epochs. The SGD algorithm is utilized for optimization and cross-entropy is employed as the loss function.

Table 1. The classification results on the CIFAR-10 and CIFAR-100 datasets.

Method	CIFAR-10	CIFAR-100
VGG-16 [11]	0.9223	0.6938
[4]	0.9578	0.7802
Ours	**0.9590**	**0.7903**

3.1 The CIFAR-10 and CIFAR-10 Datasets

The two CIFAR datasets [7] are popular for testing deep learning methods and both consist of $60,000$ images of size 32×32. Particularly, the CIFAR-10 dataset includes 10 classes, and the CIFAR-100 dataset includes 100 classes. There are $50,000$ training and $10,000$ testing images for both of them. We keep the same experimental settings as [4] on the two datasets. The results of the VGG-16 are taken from [4] and the approach in [4] is implemented by ourself. The classification results on the two datasets are listed in Table 1. As can be seen that the proposed method achieves the best performance on the two CIFAR datasets.

3.2 The Cervical Image Dataset

The cervical image dataset is provided by the Peking University First Hospital and contains cervical images of $1,700$ patients. Each image is of size 696×570.

Note that each patient in the dataset contains multiple images of three types which are generated by processing the cervix with saline, acetic acid and iodine.

According to the type of lesions of cervical cancer, the data can be divided into classes of high-grade squamous intraepithelial lesion (HSIL), low-grade squamous intra-epithelial lesion (LSIL), normal colposcopic findings (NCF), squamous cell carcinoma (SCC), squamous intra-epithelial lesion (SIL), thinprep cytology test (TCT) abnormalities, hemorrhage, neoplasms, vaginal wall lesions. Among them, the latter four types can be easily recognized by the clinicians and the SIL cases are rare in practice. Therefore, we ignore these types and only process the remaining four types. The numbers of patients in these four classes are 260, 132, 853 and 50 respectively. In particular, we can simply classify HSIL and SCC as malignant lesions and LSIL and NCF as benign lesions. Then it becomes a 2-class classification problem. For targeted treatment, it is necessary to determine the precise type of each image. In this case, it becomes a 4-class classification problem. The examples of the four categories are shown in Fig. 2.

Fig. 2. Examples of the four categories of cervical images. The categories from left to right are HSIL, SCC, LSIL and NCF.

Image Preprocessing and Augmentation. We exclude images of cervix processed with acetic acid and iodine since their amount is limited. Furthermore, we remove images with blurred content and occlusions and ensure that the number of images per patient is odd (for the convenience of majority-voting for final results). Unrelated objects in images, such as instruments, skin, hair, will impair the performance of the algorithms. Therefore, we adopt the approach in [3] to segment the cervical region to avoid interference. Then we extract image patches containing the cervical region and resize them to a size of 128 × 128.

The dataset is randomly split regarding individuals into three independent subsets for training, validation and testing with a proportion of 7:1:2. Data augmentation is performed on the training subset. For the 2-class classification problem, we obtain the mirror image of each original image and perform rotation transform with random angles (less than 20°) clockwise or counterclockwise to images from the class of malignant for 5 times. In this way, we can obtain a balanced training subset. Similarly, for the 4-class classification problem, we obtain the mirror images first and perform rotation transform to images from the classes of HSIL, LSIL and SCC for 2, 5 and 15 times respectively.

The classification results are presented in Table 2. From the results we can see that the proposed method achieves the best classification accuracy on both the 2-class and 4-class cervical image classification tasks. Compared with the VGG-16, the method in [4] can learn salient features from images with attention

Table 2. The classification results on the cervical dataset.

Method	Accuracy	Sensitivity	Spectivity	Accuracy
	2-class			4-class
VGG-16 [11]	0.7683	0.5263	0.8100	0.6589
[4]	0.7799	0.8571	0.7778	0.6705
Ours	**0.7838**	0.6667	0.7925	**0.6783**

mechanism and thus perform better. Furthermore, the proposed method utilizes multi-scale features to learn more meaningful attention maps and integrate the global feature for prediction. As a result, it achieves better results than the method in [4]. Particularly, for the 4-class problem, the class of NCF is easier to be identified due to the fact that it has more training data and the patterns of the images from this class differ a lot from other classes.

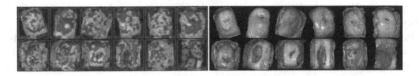

Fig. 3. The attention maps of the proposed method. The left figure presents the attention maps and the right figure present the corresponding original images.

To diagnose the type of lesion, clinicians will focus on the specific regions with lesions. Therefore, it is crucial to locate the lesions correctly. Attention mimics the visual mechanism of humans by enforcing the algorithms to focus on regions that are most related to the task. In this way, it can prevent the algorithms being interfered by unrelated factors. The intermediate attention maps of the proposed method are demonstrated in Fig. 3 and we can see that the proposed algorithm is able to focus on regions that are valuable for diagnosis.

4   Conclusions

In this paper, we propose a novel feature pyramid based attention method for cervical image classification. The method first constructs a feature pyramid by fully utilizing the multi-scale intermediate feature maps of a main network through lateral connections and feature fusion. Then the compatibility scores between the merged feature maps and the global feature are computed. Weighting merged feature maps with the compatibility scores, multi-scale attention features are extracted. Finally, the multi-scales attention features together with the global feature are concatenated and input into a fully-connected layer for final prediction. Experimental results on CIFAR datasets and cervical images collected from the Peking University First Hospital show that the proposed method achieves excellent classification performances.

Acknowledgement. This work was supported in part by the National Key Research and Development Program of China under Grant 2018YFC0910700 and in part by the National Natural Science Foundation of China under Grants 11701018, 11831002 and 81801778.

References

1. Anavi, Y., Kogan, I., Gelbart, E., Geva, O., Greenspan, H.: Visualizing and enhancing a deep learning framework using patients age and gender for chest x-ray image retrieval. In: Medical Imaging 2016: Computer-Aided Diagnosis, vol. 9785, p. 978510. International Society for Optics and Photonics (2016)
2. Ginsburg, O., et al.: The global burden of women's cancers: a grand challenge in global health. Lancet **389**(10071), 847–860 (2017)
3. Greenspan, H., et al.: Automatic detection of anatomical landmarks in uterine cervix images. IEEE Trans. Med. Imaging **28**(3), 454–468 (2009)
4. Jetley, S., Lord, N.A., Lee, N., Torr, P.H.: Learn to pay attention. arXiv preprint arXiv:1804.02391 (2018)
5. Kamnitsas, K., et al.: Efficient multi-scale 3D CNN with fully connected CRF for accurate brain lesion segmentation. Med. Image Anal. **36**, 61–78 (2017)
6. Kim, E., Huang, X.: A data driven approach to cervigram image analysis and classification. In: Celebi, M., Schaefer, G. (eds.) Color Medical Image analysis, pp. 1–13. Springer, Heidelberg (2013). https://doi.org/10.1007/978-94-007-5389-1_1
7. Krizhevsky, A., Hinton, G., et al.: Learning multiple layers of features from tiny images. Technical report, Citeseer (2009)
8. Lin, T.Y., Dollár, P., Girshick, R., He, K., Hariharan, B., Belongie, S.: Feature pyramid networks for object detection. In: Proceedings of the IEEE Conference on Computer Vision and Pattern Recognition, pp. 2117–2125 (2017)
9. Liu, Y., et al.: Detecting cancer metastases on gigapixel pathology images. arXiv preprint arXiv:1703.02442 (2017)
10. Sato, M., et al.: Application of deep learning to the classification of images from colposcopy. Oncol. Lett. **15**(3), 3518–3523 (2018)
11. Simonyan, K., Zisserman, A.: Very deep convolutional networks for large-scale image recognition. arXiv preprint arXiv:1409.1556 (2014)
12. Vaswani, A., et al.: Attention is all you need. In: Advances in Neural Information Processing Systems, pp. 5998–6008 (2017)
13. Xu, T., Zhang, H., Huang, X., Zhang, S., Metaxas, D.N.: Multimodal deep learning for cervical dysplasia diagnosis. In: Ourselin, S., Joskowicz, L., Sabuncu, M.R., Unal, G., Wells, W. (eds.) MICCAI 2016. LNCS, vol. 9901, pp. 115–123. Springer, Cham (2016). https://doi.org/10.1007/978-3-319-46723-8_14
14. Young, T., Hazarika, D., Poria, S., Cambria, E.: Recent trends in deep learning based natural language processing. IEEE Comput. Intell. Mag. **13**(3), 55–75 (2018)
15. Zhang, J., Xie, Y., Xia, Y., Shen, C.: Attention residual learning for skin lesion classification. IEEE Trans. Med. Imaging (2019)

Single-Scan Dual-Tracer Separation Network Based on Pre-trained GRU

Junyi Tong[1], Yunmei Chen[2], and Huafeng Liu[1(✉)]

[1] State Key Laboratory of Modern Optical Instrumentation, Zhejiang University,
Hangzhou 310027, China
liuhf@zju.edu.cn
[2] Department of Mathematics, University of Florida,
458 Little Hall, Gainesville, FL 32611-8105, USA
yun@ufl.edu

Abstract. In this paper, a novel network based on gated recurrent unit (GRU) is proposed for separating single-scan dual-tracer PET mixed images. Compared to conventional methods, this method can separate dual-tracer that are simultaneously injected or even labeled with the same marker, and do not require arterial blood input function. The proposed 4-layer network denoises the time activity curves (TACs) extracted from the dynamic dual-tracer reconstruction images with noise by pre-training the parameters in the first and second layer, and then uses TAC time information for dual-tracer separation. During the training stage, we optimize the network by minimizing the mean square error (MSE) objective function of the separated predicted value and ground truth. Monte Carlo is used to simulate the PET sampling environment with the mixed dual-tracer ^{62}Cu-ATSM+^{62}Cu-PTSM and ^{18}F-FDG+^{11}C-MET. Calculating the bias and variance to quantitatively analyze the results, we demonstrate that this method is more robust and better separation than the similar methods.

Keywords: PET · Dual-tracer separation · GRU

1 Introduction

Dynamic dual-tracer positron emission tomography (PET) imaging is becoming more and more important as it uses the additional information provided by the two tracers to obtain a more complete tumor status [4], which can reduce the possibility of tumor misjudgment and guide us to choose a more effective treatment plan. It can also greatly reduce the scanning time, the number of scans and the suffering of patients.

Supported in part by Shenzhen Innovation Funding (No: JCYJ20170818164343304, JCYJ20170816172431715), by the National Natural Science Foundation of China (No: U1809204, 61525106, 61427807, 61701436), and by the National Key Technology Research and Development Program of China (No: 2017YFE0104000, 2016YFC1300302).

Q. Li et al. (Eds.): MMMI 2019, LNCS 11977, pp. 43–50, 2020.
https://doi.org/10.1007/978-3-030-37969-8_6

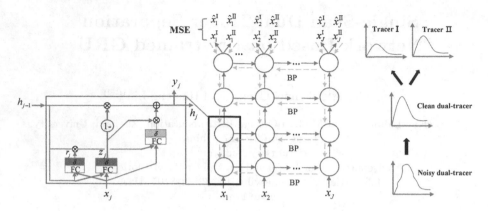

Fig. 1. Schematic diagram of dual-tracer separation network based on pre-trained GRU

The biggest challenge with dual-tracer separation is the difficulty in distinguishing photons for each tracer. Huang et al. [5] recognized that the static distribution of multiple tracers with different half-lives can be recovered from dynamic PET images based on the radioactive decay rate of the different tracers. This work laid the basic direction for the rapid separation of PET multi-tracers using dynamic imaging technology. Rust et al. [9] studied dual-tracer images of tumor hypoxia and blood flow using ^{62}Cu-ATSM and ^{62}Cu-PTSM. The method uses a dynamic single-scan with interlaced injections and treats the time overlap as the sum of effects of two tracers. The dual-tracer separation technique with instant Gammas proposed by Andreyev et al. [1] solves the separation problem from the physical level. High energy gamma rays from one of the tracers assist in signal separation.

Due to the need for alternating injections of tracers and the sampling of the arterial blood input function, traditional separation methods are invasive and require relatively long scan time. A recent work by Ruan et al. [8] using a data-driven stacked auto encoder (SAE) provides a new idea for solving this problem. The disadvantage is that the network is low in robustness.

Since the dynamic activity images of the dual-tracer contain time information, the separation of them can be analogized to speech separation. Inspired by the time-domain speech separation network [7], this paper proposes a GRU-based dual-tracer separation method, which separates mixed time signals end-to-end. Special gates can control the retention and forgetting of time information, making the separation more accurate. The PET dynamic reconstruction images with timing information are converted into TACs by pixels as inputs to the network. When training the network, the clean TACs are pre-trained firstly, which makes the network more robust to noise to some extent. The Monte Carlo simulation experiments of different phantoms and tracer pairs are used to verify the accuracy to identify tumors and robustness of the network compared with SAE.

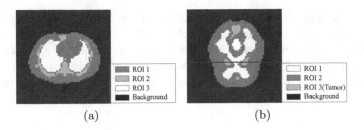

(a)   (b)

Fig. 2. (a) Zubal thorax phantom.(b) Hoffman brain phantom. Red line: lateral displacement profiles. (Color figure online)

2   Methodology

2.1   Measuring Principle of Dual-Tracer Single Acquisition

Simultaneously injecting the dual-tracer and scanning with PET for a period of time, a series of sinogram $Y_{(t)}^{dual}$ can be obtained. For sinogram, the sampling process can be expressed as follows:

$$Y_{(t)}^{dual} = GX_{(t)}^{dual} + e \tag{1}$$

$$X_{(t)}^{dual} = X_{(t)}^{I} + X_{(t)}^{II} \tag{2}$$

Combining the sinogram with system matrix G can reconstruct images, e is the noise caused by the sampling process and reconstruction algorithm. $X_{(t)}^{dual}$ mean a series of mixed reconstruction images, which can also be represented as a superposition of tracers I and II. For the mixed reconstruction images over J temporal frames, N TAC curves $X^{dual} = [x_1^{dual}, \ldots, x_i^{dual}, \ldots, x_N^{dual}]$ ($x_i^{dual} = [x_{i1}^{dual}, \ldots, x_{ij}^{dual}, \ldots, x_{iJ}^{dual}]^T$) can be obtained after removing the background. And $x_{ij}^{dual} = x_{ij}^{I} + x_{ij}^{II}$ denotes the mixed value of the jth frame of TAC i.

2.2   Dual-Tracer Separation Algorithm

The network consists of two parts as shown in Fig. 1. Part I: Input mixed noise TACs to train the first GRU layer and the second linear layer to output clean TACs; Part II: Input the trained clean TACs into the third GRU layer and the fourth linear layer to obtain the separated TACs. The formula of GRU and linear layer is as follows:

$$r_j = \sigma(W_r \cdot [h_{j-1}, x_j]) \tag{3}$$

$$z_j = \sigma(W_z \cdot [h_{j-1}, x_j]) \tag{4}$$

$$\tilde{h}_j = tanh(W_{\tilde{h}} \cdot [r_j * h_{j-1}, x_j]) \tag{5}$$

$$h_j = (1 - z_j) * h_{j-1} + z_j * \tilde{h}_j \tag{6}$$

$$y_j = W_y \cdot h_j \tag{7}$$

In the above formula, h_{j-1} is the output state of the $j-1^{th}$ frame, x_j is the input of the jth frame, and y_j is the output of the jth frame. r_j and z_j denote reset gate and update gate, respectively, for controlling the retention and forgetting of time information. At the beginning of training, we randomly initialize the output state h_0. During training, the parameters of the first and second layers are pre-trained, and then the third and fourth layers are added for overall training.

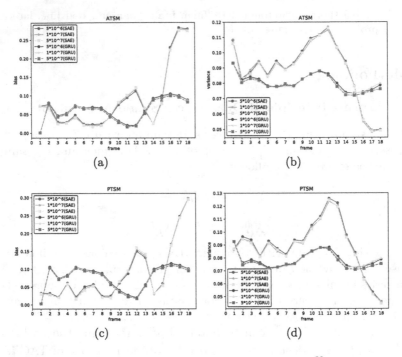

Fig. 3. Reconstruction quality analysis of each ROI. (a) Bias of ^{62}Cu-ATSM, (b) Variance of ^{62}Cu-ATSM, (c) Bias of ^{62}Cu-PTSM, (d) Variance of ^{62}Cu-PTSM.

Loss Function. The input data needs to minimize the loss function to get the optimal solution. The loss function of the GRU-based network proposed in this paper consists of two parts. The first part is the MSE of the true value $\hat{x}_i^{dual}$ and mixed predicted value x_i^{dual} to ensure that the outputs of second layer network are clean mixed TACs. The second part is the MSE of ground truth $\hat{x}_i^I, \hat{x}_i^{II}$ and network-separated TACs x_i^I, x_i^{II}. α is hyperparameter. The loss function formula is as follows:

$$Loss = \alpha \|\hat{x}_i^{dual} - x_i^{dual}\|_2^2 + \|\hat{x}_i^I - x_i^I\|_2^2 + \|\hat{x}_i^{II} - x_i^{II}\|_2^2 \qquad (8)$$

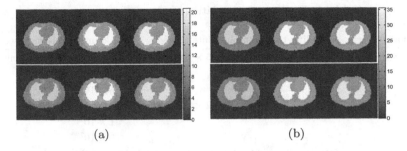

(a)   (b)

Fig. 4. Predicted activity images and ground truth of ^{62}Cu-ATSM+^{62}Cu-PTSM with 20 min sampling time and 5×10^7 counting rate. From top to bottom: ground truth and predicted results, from left to right: the 5th, 10th and 15th frames. (a) ^{62}Cu-ATSM. (b) ^{62}Cu-PTSM.

3   Experiments and Results

3.1   Data and Simulation Experiments on Phantom

In this paper, two different 64 *pixels* × 64 pixels phantoms are used to simulate the true physiological environment called Hoffman brain and Zubal thorax. As shown in Fig. 2, each phantom has three regions of interest, corresponding to different active concentrations of the tracer. Two tracer pairs: ^{62}Cu-ATSM $+^{62}$Cu-PTSM (for detecting hypoxia and blood flow) and ^{18}F-FDG $+^{11}$C-MET (for detecting tumors) were filled into the above two phantoms.

During the simulation, parallel compartment model is used to simulate the dynamic spatial distribution of a single radiotracer. The blood input function of each tracer and the kinetic parameters of the compartment models are referred to the previous paper [2,3,6,9]. Adding the 18-frame noise-free PET single-tracer images to form a mixed dynamic dual-tracer ground truth, which is used as inputs for Monte Carlo simulation. Based on the characteristics of tracers, we designed a variety of different sampling times, sampling modes and counting rates. See Table 1 for details. Using Monte Carlo (GATE) to simulate the real PET sampling process, we obtain dynamic sinograms which change with time. Because the network is difficult to deal with the sinogram directly, we use the ADMM algorithm for images reconstruction to get TACs.

All simulations were based on the geometry of the BIOGRAPH SENSATION16-HR scanner (Siemens Medical Solutions, USA), which consists of 24, 336 LSO crystals arranged in 3 rings with a diameter of 82.4 cm. The field of view is 58.5 cm on the horizontal axis and 16.2 cm in the axial direction.

3.2   Network Detail

The GRU-based network has four hidden layers, each with 32, 1, 32, 2 hidden units, respectively. After removing the background of the images, we extract the noisy TACs by pixels to input the network. The first GRU layer and the

Table 1. Dual-tracer simulation modes and parameters.

Dual tracer	Sampling time	Sampling interval	Counting rate	Dataset
ATSM+PTSM	40 min	10 × 30 s, 5 × 240 s, 3 × 300 s	5×10^6	training
			1×10^7	
			5×10^7	
	30 min	10 × 30 s, 2 × 150 s, 6 × 200 s	5×10^6	
			1×10^7	
			5×10^7	
	20 min	14 × 50 s, 2 × 100 s, 2 × 150 s	5×10^6	testing
			1×10^7	
			5×10^7	
FDG+MET	20 min	14 × 50 s, 2 × 100 s, 2 × 150 s	5×10^6	testing
			1×10^7	training
			5×10^7	
		10 × 30 s, 6 × 100 s, 2 × 150 s	5×10^6	testing
			1×10^7	training
			5×10^7	
		10 × 40 s, 5 × 70 s, 3 × 150 s	5×10^6	testing
			1×10^7	training
			5×10^7	

second linear layer are pre-trained to ensure clean TAC outputs. The separated two TACs are then output through the third GRU layer and the fourth linear layer. The primary penalty parameter α in formula (8), learning rate and the batch-size are set to 0.5, 0.002 and 32, respectively. The network optimizer is "Adam". The comparison SAE network has four hidden layers, each with 85, 70, 60 and 36 relu units. Other hyper-parameters are consistent with the proposed network to ensure the best network performance.

3.3 Result

We use bias and variance to measure the effect of dual-tracer separation. The formula is as follows:

$$bias = \frac{1}{N} \sum_{i=1}^{N} \frac{|x_i - \hat{x}_i|}{\hat{x}_i} \tag{9}$$

$$variance = \frac{1}{N} \sum_{i=1}^{N} \left(\frac{x_i - \overline{x_N}}{\hat{x}_i} \right)^2 \tag{10}$$

Where $\hat{x}_i$ is the ground truth of the ith pixel of a ROI and x_i means the predicted value of the ith pixel in the ROI. N and $\overline{x_N}$ represent the total number of pixels and the average predicted value in one ROI, respectively.

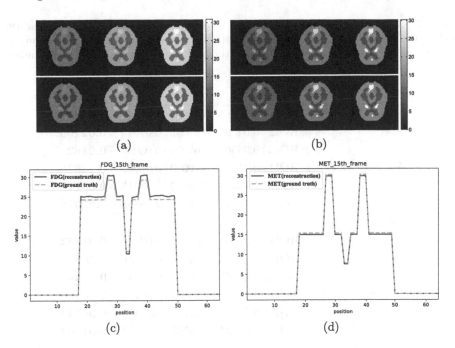

Fig. 5. Predicted activity images and ground truth of ^{18}F-FDG+^{11}C-MET with 10 × 40s, 5 × 70s, 3 × 150s sampling interval and 5 × 10^7 counting rate. From top to bottom: ground truth and predicted results, from left to right: the 5th, 10th and 15th frames. (a) ^{18}F-FDG, (b) ^{11}C-MET, (c) Profiles of 15th ^{18}F-FDG image, (d) Profiles of 15th ^{11}C-MET image.

^{62}Cu-ATSM+^{62}Cu-PTSM. Figure 3 shows that the higher the counting rate, the bias and variance of the separation results are only slightly reduced, which means the network is robust to the counting rate. And comparing with SAE, the proposed network has smoother bias and variance. Since the half-life of Cu is 9.7 min, the tracer concentration first increases and then decreases from the first frame to the 18th frame, and the corresponding bias is just the opposite. By the 12th frame, the concentration is the largest and the bias is reduced to the minimum. Figure 4 shows the results of the tracer separation, and it can be seen that the image boundaries are clear.

^{18}F-FDG+^{11}C-MET. Table 2 shows that the separation results of two tracers are very good. Because the half-lives of C and F are 20.4 min and 110 min, respectively, during the 20-minute sampling period, the bias of two tracers has been decreasing. Figure 5 indicates that comparing with FDG, MET is more clearly engraved on the tumor area, and can be used clinically to assist FDG tracer to achieve rapid sampling diagnosis of tumor. The boundary between the tumor and normal tissue is very clear, and the separation results of the network are very good.

Table 2. The bias and variance of dual-tracer (^{18}F-FDG $+^{11}$C-MET) separation results (the 4th, 8th, 12th, 16th frames) with 10×40 s, 5×70 s, 3×150 s sampling interval and 5×10^6 counting rate. Besides, the best results are shown in bold numbers.

FDG					
Bias	ROI1	0.12988	0.13299	0.10979	**0.04407**
	ROI2	0.04566	0.04062	0.02929	**0.02105**
	ROI3	0.08563	0.05276	0.03053	**0.0182**
Variance ($\times 10^{-4}$)	ROI1	0.31623	0.70030	0.56002	**0.24675**
	ROI2	1.10529	2.79649	1.68313	**0.68719**
	ROI3	2.12170	5.55187	2.58368	**0.97379**
MET					
Bias	ROI1	0.02058	0.03340	0.03266	**0.01952**
	ROI2	0.01052	0.01633	**0.01712**	0.02710
	ROI3	0.11763	0.06729	0.02648	**0.02494**
Variance ($\times 10^{-4}$)	ROI1	0.14207	0.23012	0.03793	**0.00560**
	ROI2	0.32096	0.61241	0.09451	**0.00946**
	ROI3	0.79734	1.46765	0.17941	**0.01262**

References

1. Andreyev, A., Celler, A.: Dual-isotope PET using positron-gamma emitters. Phys. Med. Biol. **56**(14), 4539 (2011)
2. Cheng, X., et al.: Direct parametric image reconstruction in reduced parameter space for rapid multi-tracer PET imaging. IEEE Trans. Med. Imaging **34**(7), 1498–1512 (2015)
3. Feng, D., Wong, K.P., Wu, C.M., Siu, W.C.: A technique for extracting physiological parameters and the required input function simultaneously from PET image measurements: theory and simulation study. IEEE Trans. Inf Technol. Biomed. **1**(4), 243–254 (1997)
4. Guo, J., et al.: 18F-Alfatide II and 18F-FDG dual-tracer dynamic PET for parametric, early prediction of tumor response to therapy. J. Nucl. Med. **55**(1), 154–160 (2014)
5. Huang, S., Carson, A., Hoffman, E., Phelps, D.: An investigation of a double-tracer technique for positron computerized tomography. J. Nucl. Med. **23**, 816–822 (1982)
6. Kadrmas, D.J., Rust, T.C.: Feasibility of rapid multitracer PET tumor imaging. IEEE Trans. Nucl. Sci. **52**(5), 1341–1347 (2005)
7. Luo, Y., Mesgarani, N.: TasNet: time-domain audio separation network for real-time, single-channel speech separation. In: 2018 IEEE International Conference on Acoustics, Speech and Signal Processing (ICASSP), pp. 696–700. IEEE (2018)
8. Ruan, D., Liu, H.: Separation of a mixture of simultaneous dual-tracer PET signals: a data-driven approach. IEEE Trans. Nucl. Sci. **64**(9), 2588–2597 (2017)
9. Rust, T., Kadrmas, D.: Rapid dual-tracer PTSM+ ATSM PET imaging of tumour blood flow and hypoxia: a simulation study. Phys. Med. Biol. **51**(1), 61 (2005)

PGU-net+: Progressive Growing of U-net+ for Automated Cervical Nuclei Segmentation

Jie Zhao[1], Lei Dai[2], Mo Zhang[2], Fei Yu[2], Meng Li[2], Hongfeng Li[1], Wenjia Wang[2], and Li Zhang[1(✉)]

[1] Center for Data Science in Health and Medicine,
Peking University, Beijing 100871, China
zhangli_pku@pku.edu.cn
[2] Center for Data Science, Peking University, Beijing 100871, China

Abstract. Automated cervical nucleus segmentation based on deep learning can effectively improve the quantitative analysis of cervical cancer. However, accurate nuclei segmentation is still challenging. The classic U-net has not achieved satisfactory results on this task, because it mixes the information of different scales that affect each other, which limits the segmentation accuracy of the model. To solve this problem, we propose a progressive growing U-net (PGU-net+) model, which uses two paradigms to extract image features at different scales in a more independent way. First, we add residual modules between different scales of U-net, which enforces the model to learn the approximate shape of the annotation in the coarser scale, and to learn the residual between the annotation and the approximate shape in the finer scale. Second, we start to train the model with the coarsest part and then progressively add finer part to the training until the full model is included. When we train a finer part, we will reduce the learning rate of the previous coarser part, which further ensures that the model independently extracts information from different scales. We conduct several comparative experiments on the Herlev dataset. The experimental results show that the PGU-net+ has superior accuracy than the previous state-of-the-art methods on cervical nuclei segmentation.

Keywords: Cervical nuclei segmentation · Pap smear test · Multi-scale · Progressive growing · Residual module

1 Introduction

Pap smear is an important test for early screening of precancerous lesions and malignant tumors in gynecology. Accurate segmentation of cervical cancer cells, especially the segmentation of the nuclei, is significant to quantitatively analyze the cervical cancer. Traditional cervical segmentation methods based on image

---

J. Zhao and L. Dai—Joint First Authors.

© Springer Nature Switzerland AG 2020
Q. Li et al. (Eds.): MMMI 2019, LNCS 11977, pp. 51–58, 2020.
https://doi.org/10.1007/978-3-030-37969-8_7

representation are widely used, such as Wavelet [1], support vector machines [2], template fitting [3], adaptive thresholding [4], genetic algorithms [5] and graph-cuts [6]. Such methods are based on low-level hand-crafted features that usually represent the texture features of the image rather than high-level semantic features. Since the cervical cells of different disease stages undergo global (semantic) changes, if these methods are unable to effectively extract the semantic information of the images, their segmentation accuracy will not satisfy the actual clinical requirements.

The method of deep learning pixel-based object segmentation or detection can simultaneously take into account the characteristic information of different cell structures. The structure of a neural network adjusts the sizes of the receptive fields to adapt to different sizes of targets. Continuous feature extraction through multiple iterations can greatly promote the accuracy of segmentation results. Traditional convolutional neural network U-net [7] realizes multi-scale information extraction through skip connection. The multi-scale information may have much redundancy and repetition. The use of fixed-size receptive fields for different scale targets is limited to multi-scale learning. Many studies have begun to focus on multi-scale information extraction methods for different target sizes and shapes, such as increasing the receptive field, adding dilated convolution, and merging feature information of different convolution layers, thus improving the classification accuracy of each pixel and generalization of detail features. [8] proposed multi-scale convolutional networks and segmentation methods for cervical nucleus and cytoplasm based on graph partitioning. Song et al. uses a multi-scale deep convolutional neural network to extract diverse feature information and segment overlapping cervical cells [9]. The dilated convolution model, which combines multi-scale context information while maintains the receptive field of the original network without losing the resolution of the image space. It has good effects in image classification, target detection and semantic segmentation [10,11]. However, the dilated rate of the dilated convolution is difficult to design. The artificially designed dilated convolution cannot take into account the characteristic information embodied by the targets of different sizes and shapes. At the same time, learning the feature information of different scales is powerless for the neural network.

To address the aforementioned problems, we propose a novel model - the progressive growing of U-net with residual modules (PGU-net+). Based on the classic U-net, we propose two improvements in the network architecture. First, we added residual modules between different stages (i.e. scales) of the classic U-net. In the first stage with the lowest resolution, we downsample the image and the annotation and train the coarsest part of the model, which learns an approximate shape of the segmentation. We then pass this approximate shape through a residual connection to the next stage with higher resolution, which only learns the residuals of the approximate shape and the annotation (images and annotations will be resampled accordingly in all stages). Thus at each stage, we enforce the model to learn the information related to the current scale. We name this architecture as U-net+. Experiments show that U-net+ can effectively improve the segmentation accuracies.

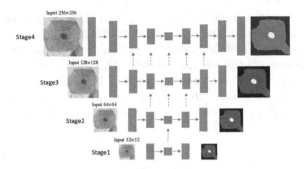

Fig. 1. Flowchart of the experimental procedure. We first use the low-resolution (32×32) image as the input of Stage1, and perform the convolution operation through the solid arrow to get the feature map of each layer (blue boxes). Then we progressively increase the resolution of the input image (the second stage is 64×64, the third stage is 128×128, and the fourth stage is 256×256) and the network is deepened to obtain output results of different sizes. During the stage1 to stage4 process, the middle-layer parameters of the previous stage are continuously transferred (by the dashed arrow). (Color figure online)

Second, we adopt a network training paradigm in [12], called progressive growing. We start to train the model with the coarsest part with downsampled images and annotations, and then progressively add finer part to the training until the full model is included. When training a finer part, we will reduce the learning rate of the previous coarser part, which further encourages the model to extract information from different scales independently. In addition, such paradigm significantly reduces the computational consumption than training the entire model simultaneously. Figure 1 shows the flow chart of this method comprises four stages.

2 Method

Classical U-net comprises two major parts: contracting path and expansive path. In the contracting path of deep neural networks, a series of convolution operations can extract feature information to generate coarser feature maps. In the expansive path, corresponding decoding stages progressively recover the resolution of feature maps from coarse to fine.

2.1 Residual Module

In order to avoid information loss, we introduce a residual module (as shown in Fig. 2) between adjacent scales. The low-resolution feature map of the previous layer is added directly to the high-resolution feature map of the next layer at the pixel level to form residual module. The module is defined as follows:

$$y(p) = F(X(p), W_2(p)) + G(X(p), W_2(p)) \tag{1}$$

Here $X(p)$ represents the input feature map, $W_1(p)$, $W_2(p)$ denote the weight of the convolution kernel, $y(p)$ represents the output feature map, and the function $F(x, w)$ is the convolution of the expansive path and the maximum pooling operation. $G(x, w)$ represents the residual module. This kind of structure can extract more abundant multi-scale information without increasing the parameters and calculation cost. At each stage, the current network pays more attention to the residual information of adjacent scales to ensure good performance.

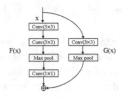

Fig. 2. Residual module.

2.2   (Progressive Growing) PG Method

Traditional convolution kernels or deformable convolution kernels simultaneously learn target information of all scales, which can easily lead to a large number of repetitive or redundant features. If the network is deepened and widened, it will result in high computational and memory cost. Our proposed PGU-net+ model extracts multi-scale feature by introducing a progressive growing [12] training approach. As shown in Fig. 1, we set up 4 training phases. In the first phase, we input a low-resolution image (32×32) to a small U-net network to get the same size of low-resolution output. Then we gradually increase the resolution of the input image to 64×64, 128×128 and 256×256, and continuously add convolution layers to the network to form deeper U-net structures. This type of training allows the network to learn large-scale image coarse structure information first, and then focus on more detailed features at a later stage, rather than learn information of all the scale at the same time. At each stage, the model receives input images of different sizes, so that multi-scale information of target regions of different sizes can be learned step by step. This method makes the model converge faster and have better generalization ability and stability without extra parameters and calculations. Figure 3 shows the U-net structure in the final stage with the residual module added to each expansive path.

We introduce residual module in the extended path of the U-net structure, and adopt a progressive growing training method. At each stage, the model iteratively learns the residual information of adjacent scales. All existing layers in networks remain trainable throughout the training process. When new layers are added to the networks, we adjust smaller learning rate to well-trained, smaller-resolution layers with transferred parameters to avoid sudden shocks on existing networks. By migrating low-resolution image features, the learning of

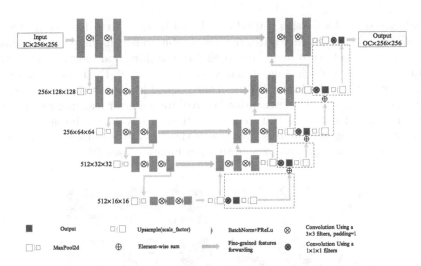

Fig. 3. The U-net structure in the final stage. By migrating the third stage intermediate layer and adding a layer of upsampling and downsampling operations to form the final model structure. In the expansive path, the residual information of the adjacent scale is specially learned, and the input image and the output result size are both 256 × 256.

high-resolution images is easier, and the convergence process is faster. The task division of multi-scale learning is further clarified, and the extracted multi-scale information is more accurate and rich.

3   Experiment and Result

3.1   Data Description

In response to our proposed PGU-net+ structure, this experiment validates our method on the Herlev dataset. The dataset contains 917 images of cervical cancer cells, with each image containing four parts: background, cytoplasm, nucleus and unknown area. Here, we manually determine the unknown area as the background. Considering the difference between large and small nuclei, large and small nuclei are segmented as two types during model training, and all images are normalized to zero mean with unit variance intensity and are resized to a size of 256 × 256.

3.2   Implementation Details

We train the model on a single NVIDIA GPU-TITAN. In the first stage, a 32 × 32 raw data is used as input for a small U-net. In the expansive path, the low-resolution feature map is directly doubled and then added to the adjacent high-resolution output to form a residual module, so that the network focuses on learning the residual information of different scales. In the second stage, the original image of 64 × 64 size is used as a U-net input with 2 downsampling and upsampling. In the expansive path, the low resolution feature map is also doubled and added to the adjacent one. And so on into the third and fourth stages. After training 40 epochs at each stage, the next stage is entered. During the parameter transferring process, the learning rate of the trained low-resolution convolutional layer is set to 1e-6, and the newly added convolutional layer learning rate is set to 1e-4 to maintain large-scale feature information and avoid the impact of model changes on existing parameters. We use RMSprop optimization to adaptively adjust the model weights, and the activation function uses RELU.

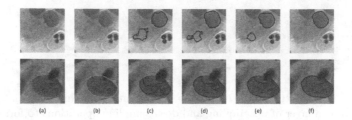

Fig. 4. Examples of the segmentation results. (a) Pap smear images, (b) Manual annotations, (c) Segmentation results of U-net, (d) Segmentation results of U-net+, (e) Segmentation results of PGU-net, (f) Segmentation results of PGU-net+.

Table 1. Four sets of experimental results (classical U-net, U-net+, PGU-net, and pgU-net+).

Methods	U-net	U-Net+	PGU-net	**PGU-net+**
ZSI	0.879 ± 0.14	0.907 ± 0.10	0.911 ± 0.10	**0.926 ± 0.09**
Precision	0.857 ± 0.19	0.878 ± 0.14	0.890 ± 0.12	**0.901 ± 0.13**
Recall	0.941 ± 0.08	0.960 ± 0.07	0.950 ± 0.11	**0.968 ± 0.04**

3.3 Experimental Results

We conduct four sets of experiments. The first group uses a traditional U-net structure to perform nuclear segmentation on 256 × 256 images, including four layers of downsampling and upsampling operations. The second group (short for U-net+) adds a residual module to the expansive path of the traditional U-net structure, making it easier for the training process to grasp features at different scales. The third group (short for PGU-net) applies the progressive growing training method to the traditional U-net structure, continuously increasing the resolution of the input image from 32 to 256 and slowly migrating the low-resolution layer parameters trained in the previous stage. The fourth group adds residual modules in the traditional U-net structure and introduces a progressive growing training mode. The superiority of our proposed PGU-net+ is verified by comparing the four sets of experiments.

By comparing experiments on the Herlev dataset, a total of four set of segmentation results for the dataset are summarized. As shown in Table 1, we give three indicators of ZSI, precision and recall. It shows that the U-net network structure with residual module (PGU-net+ model and U-net+ model) is superior to the classic U-net neural network (PGU-net model and U-net model). The progressive growing U-net network structure (PGU-net+ model and PGU-net model) is superior to the classic U-net neural network (U-net+ model and U-net model). The progressive growing with the residual module U-net structure we proposed achieves the best segmentation results. The results of the two groups of cell segmentation experiments are shown in Fig. 4. It can be seen that the PGU-net+ has better segmentation results for cells of different sizes and shapes. We also compare other studies for this dataset. Table 2 shows the superiority of our model in the three indicators of ZSI, precision and recall under a single model. Our proposed PGU-net+ structure has a segmentation accuracy of 0.925 on the Herlev dataset, and the parameter amount (13M) and computation are much smaller than other models.

Table 2. Comparison of the state-of-the-art methods and proposed method

Method	ZSI	Precision	Recall
Unsupervised [13]	0.89 ± 0.15	0.88 ± 0.15	0.93 ± 0.15
FCM [14]	0.80 ± 0.24	0.85 ± 0.21	0.83 ± 0.25
SP-CNN [15]	0.90	0.89	0.91
DenseUnet [16]	0.91 ± 0.12	0.893 ± 0.14	0.956 ± 0.08
Our Method	$\mathbf{0.925 \pm 0.09}$	$\mathbf{0.901 \pm 0.13}$	$\mathbf{0.968 \pm 0.04}$

4 Conclusion

In this work, we propose to add the residual module in the expansive path of the classic U-net structure, and adopt the progressive growing training mode. Four models (PGU-net+, U-net+, PGU-net and U-net) are used to test on the Herlev dataset. The experimental results show that our model is effective to extract multi-scale information, making the task of extracting multi-scale information more explicit. Furthermore, this residual module can be easily inserted into other higher-order and more complex neural network structures, and the progressive growing training method can also be optimized to solve different scale target detection and target segmentation problems in other fields.

Acknowledgments. This work is supported in part by the National Key Research and Development Program of China under Grant 2018YFC0910700 and the National Natural Science Foundation of China (NSFC) under Grants 81801778, 11831002, 11701018.

References

1. Bora, K., Chowdhury, M., Mahanta, L.B., Kundu, M.K., Das, A.K.: Automated classification of pap smear images to detect cervical dysplasia. Comput. Methods Programs Biomed. **138**, 31–47 (2017)
2. Tareef, A., Song, Y., Cai, W., Feng, D.D., Chen, M.: Automated three-stage nucleus and cytoplasm segmentation of overlapping cells. In: 2014 13th International Conference on Control Automation Robotics & Vision (ICARCV), pp. 865–870. IEEE (2014)
3. Hai-Shan, W., Barba, J., Gil, J.: A parametric fitting algorithm for segmentation of cell images. IEEE Trans. Biomed. Eng. **45**(3), 400–407 (1998)
4. Plissiti, M.E., Nikou, C., Charchanti, A.: Automated detection of cell nuclei in pap smear images using morphological reconstruction and clustering. IEEE Trans. Inf Technol. Biomed. **15**(2), 233–241 (2010)
5. Lassouaoui, N., Hamami, L.: Genetic algorithms and multifractal segmentation of cervical cell images. In: Proceedings of Seventh International Symposium on Signal Processing and its Applications, vol. 2, pp. 1–4. IEEE (2003)
6. Zhang, L., et al.: Segmentation of cytoplasm and nuclei of abnormal cells in cervical cytology using global and local graph cuts. Comput. Med. Imaging Graph. **38**(5), 369–380 (2014)
7. Ronneberger, O., Fischer, P., Brox, T.: U-Net: convolutional networks for biomedical image segmentation. In: Navab, N., Hornegger, J., Wells, W.M., Frangi, A.F. (eds.) MICCAI 2015. LNCS, vol. 9351, pp. 234–241. Springer, Cham (2015). https://doi.org/10.1007/978-3-319-24574-4_28
8. Song, Y., Zhang, L., Chen, S., Ni, D., Lei, B., Wang, T.: Accurate segmentation of cervical cytoplasm and nuclei based on multiscale convolutional network and graph partitioning. IEEE Trans. Biomed. Eng. **62**(10), 2421–2433 (2015)
9. Song, Y., et al.: Accurate cervical cell segmentation from overlapping clumps in pap smear images. IEEE Trans. Med. Imaging **36**(1), 288–300 (2016)
10. Yu, F., Koltun, V., Funkhouser, T.: Dilated residual networks. In: Proceedings of the IEEE Conference on Computer Vision and Pattern Recognition, pp. 472–480 (2017)
11. Chen, L.-C., Papandreou, G., Kokkinos, I., Murphy, K., Yuille, A.L.: DeepLab: semantic image segmentation with deep convolutional nets, atrous convolution, and fully connected CRFs. IEEE Trans. Pattern Anal. Mach. Intell. **40**(4), 834–848 (2017)
12. Karras, T., Aila, T., Laine, S., Lehtinen, J.: Progressive growing of gans for improved quality, stability, and variation. arXiv preprint arXiv:1710.10196 (2017)
13. GençTav, A., Aksoy, S., ÖNder, S.: Unsupervised segmentation and classification of cervical cell images. Pattern Recognit. **45**(12), 4151–4168 (2012)
14. Chankong, T., Theera-Umpon, N., Auephanwiriyakul, S.: Automatic cervical cell segmentation and classification in pap smears. Comput. Methods Programs Biomed. **113**(2), 539–556 (2014)
15. Gautam, S., Bhavsar, A., Sao, A.K., Harinarayan, K.K.: CNN based segmentation of nuclei in pap-smear images with selective pre-processing. In: Medical Imaging 2018: Digital Pathology, vol. 10581, p. 105810X. International Society for Optics and Photonics (2018)
16. Zhao, J., Li, Q., Li, X., Li, H., Zhang, L.: Automated segmentation of cervical nuclei in pap smear images using deformable multi-path ensemble model. In: 2019 IEEE 16th International Symposium on Biomedical Imaging (ISBI 2019), pp. 1514–1518. IEEE (2019)

Automated Classification of Arterioles and Venules for Retina Fundus Images Using Dual Deeply-Supervised Network

Meng Li[1], Jie Zhao[2], Yan Zhang[3], Danmei He[3], Jinqiong Zhou[4], Jia Jia[3], Haicheng She[4], Quanzheng Li[2,5], and Li Zhang[1,2(✉)]

[1] Center for Data Science, Peking University, Beijing 100871, China
zhangli_pku@pku.edu.cn
[2] Beijing Institute of Big Data Research, Peking University, Beijing 100871, China
[3] Department of Cardiology, Peking University First Hospital, Beijing 100034, China
[4] Beijing Tongren Hospital, Capital Medical University, Beijing 100730, China
[5] MGH/BWH Center for Clinical Data Science, Boston, MA 02115, USA

Abstract. Different patterns of retinal arterioles and venules in the fundus images form an important metric to measure the disease severity. Therefore, an accurate classification of arterioles and venules is greatly necessary. In this work, we propose a novel network, named as the dual Deeply-Supervised Network (dual DSN), to classify arterioles and venules on retinal fundus images. We employ the U-shape network (U-Net) as the backbone of our proposed model. Our proposed dual DSN produces an auxiliary output of the network at every scale, which generates a loss by comparing to the manual annotation. The losses in the encoding path of dual DSN regularize the low-level features, while those in the decoding path of dual DSN regularize the high-level features. In sum, such losses in dual DSN form dual supervision to the backbone U-Net and capture the multi-level features of the input image, which improves the classification of retinal arterioles and venules. The experimental results demonstrate that the proposed dual DSN outperforms the previous state-of-the-art methods on DRIVE dataset with an accuracy of 95.0%.

Keywords: Deep learning · Convolution neural network · Dual supervision · Skip connection

1 Introduction

The changes of retinal vasculature contain substantial diagnostic information for many vascular and systematic diseases. Specifically, diseases may affect arterioles and venules differently. For example, in hypertensive patients, the width of arterioles usually shrinks faster than that of venules, while in diabetic patients, we usually observe the expansion of venules first. Therefore, accurate classification of retinal arterioles and venules in retinal fundus images has a great potential to improve the early diagnosis and treatment of these diseases.

© Springer Nature Switzerland AG 2020
Q. Li et al. (Eds.): MMMI 2019, LNCS 11977, pp. 59–67, 2020.
https://doi.org/10.1007/978-3-030-37969-8_8

Arteriovenous classification methods can be divided into three types: feature-based methods, graph-based methods, and deep learning based methods. For the feature-based methods, multiple image features are extracted to distinguish the arterioles and the venules. Kondermann et al. [4] proposed to use vessel intensity and central light reflex to identify the arterioles, because of the high hemoglobin levels in arterial blood. Xu et al. [9] proposed to use first-order and second-order texture features to capture the discriminating characteristics of arterioles and venules. For the graph-based methods, Hu et al. [3] proposed to use a graph-based meta-heuristic algorithm to separate the arterioles trees from the venules trees and Estrada et al. [2] proposed to use the underlying vessel topology to form a graph model which optimizes the separation of arteriole and venule trees.

For the deep learning based methods, a number of supervised learning approaches are reported for the arteriovenous classification. For example, Albadawi et al. [1] reported using fully convolutional network to segment the retinal vessels. Welikala et al. [8] presented a two-stage method. Retinal vessel centerlines are first obtained and centerlines are then classified as arterioles or venules by a 6-layer neural network. Although the aforementioned methods have achieved high accuracy, segmentation errors often occur in identifying small vessel segments. One possible explanation is that these deep learning based methods cannot effectively extract the low-level and high-level features simultaneously.

To solve this issue, we propose to use dual Deeply-Supervised Network (dual DSN) to classify arterioles and venules on retinal fundus images. We adapt U-Net [5] as our network backbone which contains an encoding path to extract low-level features and a decoding path to extract high-level features. We then add supervision to regularize the information extraction on every scale of the network. More specifically, the proposed model aggregates each feature map of the network into an auxiliary output, an auxiliary loss is then computed based on the difference between the auxiliary output and the manual annotation. The auxiliary losses in the encoding path ensure to obtain optimal low-level features and those in the decoding path guarantee to extract optimal high-level features. We name such mechanism as dual supervision. The highest classification accuracy proves that our proposed dual DSN outperforms the previous state-of-the-art methods on DRIVE dataset [7]. The experimental results demonstrate that the dual supervision with losses on all scales is most effective to classify the arterioles and the venules on retinal fundus images.

2   Methods

2.1   Dual Supervision

U-Net [5] is a fully convolution network with encoder-decoder structure that widely used in semantic segmentation tasks. During the training of the U-Net, low-level features extracted by the encoder are connected to high-level features extracted by the decoder via skip connections. However, plain skip connections may confound the features from different levels and decrease the accuracy of the

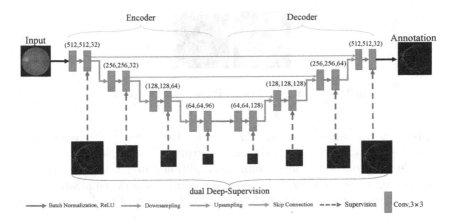

Fig. 1. The overall architecture of dual DSN.

model. To solve this issue, UNet++ [13] proposed to combine the multi-level features via densely nested connections with deep supervision, arguing that a deeply supervised structure is preferable to deal with multi-level feature extraction.

Motivated by such observation, we propose a dual Deeply-Supervised Network (dual DSN) by adding separate supervision on the every feature map of the encoder module and the decoder module simultaneously. The supervision applied on each feature map is defined by the following procedures.

Assuming the feature maps in the encoder-decoder structure with skip connections are:

$$E_i, D_j . i = 1, 2, \cdots, m, j = 1, 2, \cdots, n. \tag{1}$$

where E_i represents the ith feature map in downsampling stages from the encoder (from top to bottom), D_j represents the jth feature map in upsampling stages from the decoder (from bottom to top). For each feature map above, we use k bottleneck convolutions with kernel size $1 * 1$ to create an auxiliary output, while k is the number of the categories of the muiti-class task. We get $m + n$ auxiliary outputs each with k channels for all the feature maps of all the scales in this way, which are defined as:

$$W = \{E_{i,k}, D_{j,k} . i = 1, 2, \cdots, m; j = 1, 2, \cdots, n\}. \tag{2}$$

A bilinear interpolation is used to expand the sizes of the auxiliary outputs in W to the original image size if necessary. We then employ the softmax operation on each auxiliary output in W, and Cross-Entropy loss is computed between each auxiliary output and the manual annotation Q. To emphasize the effect of the last layer of the decoder (also the final output $output_{final}$), we introduce an additional loss at this layer, L_{last}. In sum, there are $m + n + 1$ loss functions in total, which form the final objective function of the proposed method:

$$L = \sum_{i=1}^{m} CE(E_{i,k}, Q) + \sum_{j=1}^{n} CE(D_{j,k}, Q) + L_{last}. \tag{3}$$

Fig. 2. Left: original image. Right: ananotated image.

CE represents Cross-Entropy loss, $L_{last} = CE(output_{final}, Q)$ represents the Cross-Entropy loss of the last layer (final output). The U-Net only minimizes L_{last}, while the proposed model minimizes additional $m + n$ losses to regularize the feature extraction process throughout the entire network. Dual supervision is the main difference compared with UNet++ and can avoid vanishing gradients.

2.2 Network Architecture

The proposed network for classification of arterioles and venules is based on the U-Net. In the encoder module, three convolution layers followed by three pooling operations are applied. In each convolution layer, two 3×3 convolutions followed by a batch normalization and a rectified linear unit (ReLU). As the classification of arterioles and venules in pixel-level is a densely predicted task, we replace maximum pooling operations with a concatenation of convolutions and average pooling operation for more fine details. In the decoder module, the network includes three upsampling operations. Each upsampling operation is followed by two convolutions. The proposed model has 8 convolutional layers in total. We compute auxiliary losses after each convolutional block of the network, by adding an extra loss on the final layer, there are 9 auxiliary outputs and 9 losses in total. The overall architecture of the proposed network is shown in Fig. 1

3 Experiments and Results

3.1 Data

Data Description: We conduct experiment on a public dataset, the DRIVE dataset. The dataset contains 40 color retinal fundus images with the annotations of arterioles, venules, intersections of arterioles and venules, uncertain and background in pixel-level by doctors. An example of retinal fundus images and annotations is shown in Fig. 2. The image size is 584*565. We use a randomly selected subset of 30 images for training and the rest 10 images for testing.

Data Augmentation: We use data augmentation to enlarge the 30 training images to reduce overfitting. The images are firstly randomly cropped to a size of 512*512, and random scaling in [0.5, 1.5], horizontal and vertical flipping, width and height shifting, and random clipping are then applied to augment the dataset. Finally, the size of training dataset is expended to 2,490 images. We use the first of 80% augmented data for training, the last 20% for validation, the rest 10 images of the original 40 images for testing.

Pre-processing: We use median adjust strategy as shown in [9], and then the images are normalized to [0, 1] by dividing 255. A case-level five-fold cross validation is performed within the training dataset.

As shown in Fig. 2, five labels are provided by the annotations which red for arterioles, blue for venules, green for intersections, black for background and white for uncertain pixels. In the experiment, we remove uncertain pixels for simplicity. Furthermore, the intersections have textural characteristics from both arterioles or venules, we set a very small class weight $(1e-6)$ to the intersections to avoid ambiguity during training, and they are actually classified as arterioles or venules in our model, this is in line with reality. With the processing above, our model is actually training a three-class classification of every pixel in the entire retina fundus image, it assigns every pixel in the entire retina fundus image as arteriole, venule or background.

3.2 Prevent Overfitting

Though we have increased the size of the training set 83 times though data augmentation, to prevent overfitting for training such a small dataset, we reduce the parameters of the model by using as few convolutions filters as possible. We use 32,32,64,96,128 convolution filters (from top to bottom) in the encoder stage, and 128,64,32 (from bottom to top) convolution filters in the decoder stage, respectively. The proposed model has 0.9 million parameters in total. We use dropout and add batch normalization after each convolution layer in the neural network.

3.3 Training and Experimental Results

The proposed model is trained by standard backpropagation and stochastic gradient descent (SGD) with momentum 0.9. The initial learning rate is $1e-4$. We train the model with the batch size of 4 on two NVIDA 1080ti GPUs for 1200 epochs, and thus 597,600 iterations.

We take the evaluation strategy as described in Xu et al. [11], since they also classify the categories of all the pixels in the entire blood vessels, not only the categories of centerline pixels. Evaluation strategy in Xu et al. [11] is a two-stage strategy. First, vessel segmentation result is evaluated, in this stage arterioles

Table 1. Vessel segmentation results compared with others.

Methods	Year	Se	Sp	Acc	AUC
Wang et al. [6]	2015	0.817	0.973	0.977	0.948
Xu et al. [10]	2016	0.786	0.955	0.933	0.959
Zhang et al. [12]	2016	0.774	0.973	0.948	0.964
Xu et al. [11]	2018	0.944	0.955	0.954	0.987
dual DSN	**2019**	**0.805**	**0.989**	**0.969**	**0.988**

Table 2. Different supervision settings.

Model	Accuracy	Description
U-Net	92.7%	Evaluated on the whole detected vessels
Encoder-Supervised U-Net	94.0%	Evaluated on the whole detected vessels
Decoder-Supervised U-Net	94.4%	Evaluated on the whole detected vessels
dual DSN	**95.0%**	**Evaluated on the whole detected vessels**

Fig. 3. Classification results compared with U-Net. From left to right: original image, annotated image, result using U-Net, result using dual DSN. (best in room)

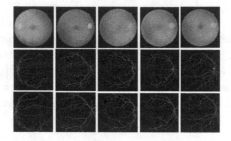

Fig. 4. Classification examples. Top: original image. Middle: annotated image. Bottom: classification result using dual DSN. Red represents arterioles and blue represents venules. Uncertain pixels (white) are not classified, and the intersections (green) are classified as arterioles or venules due to a small class weight. (best in room) (Color figure online)

and venules are all considered as vessels. Second, accuracy of classification of arterioles and venules is evaluated on all the vessel pixels that have been classified correct, thus evaluated on detected vessels. We refer readers to Xu et al. [11] for more details.

For the first-stage evaluation, sensitivity (Se), specificity (Sp), overall accuracy (Acc), and Area under the ROC curve (AUC) are used to evaluate the performance of the vessel segmentation:

$$Se = \frac{TP}{TP + FN}, Sp = \frac{TN}{TN + FP}, Acc = \frac{TP + TN}{TP + FP + TN + FN} \qquad (4)$$

where the positive class resperents for the vessels (regardless of arterioles or venules), the negative class resperents for the background. Vessel segmentation results compared with others are show in Table 1.

Table 3. Results compared with others.

Methods	Year	Accuracy	Description
Hu et al. [3]	2013	84%	Evaluated on all vessel centerline locations
Estrada et al. [2]	2015	93.5%	Evaluated on known vessel centerline locations
Welikala et al. [8]	2017	91.97%	Evaluated on known vessel centerline locations
Xu et al. [9]	2017	92.3%	Evaluated on the correctly detected vessels
AlBadaWa et al. [1]	2018	93.5%	Evaluated on known vessel centerline locations
Xu et al. [11]	2018	90.0%	Evaluated on the whole detected vessels
dual DSN	**2019**	**95.0%**	**Evaluated on the whole detected vessels**

Compared with Xu et al. [11], our proposed dual DSN achieves comparable results in Sp, Acc and AUC. The Se of dual DSN is 0.805, which is comparable with other high performance methods except Xu et al. [11], which is 0.944 and much higher than other methods. The reason is that Xu et al. [11] uses class weight in their loss function, which using 10 for arteriole, 5 for venule, and 1 for background. For fair comparison, we also conduct experiment that using class weight which 10 for arteriole, 5 for venule, and 1 for background. We found that Se varies from 0.805 to 0.939 while other indicators almost unchanged, more tiny vessels are segmented from the background. But the vessels predicted are much wider than the normal vessels. Using class weight more than 1 will cause false positive vessels, this false positive phenomenon is not truthfulness and will cause more misclassification of arterioles and venules, so we don't use class weight.

For the second-stage evaluation, $Accuracy$ is used to evaluate the performance of the model on classifying arterioles and venules on the detected vessels:

$$FPR_{at} = \frac{FP_{at}}{TP_{at} + FP_{at}}, FPR_{ve} = \frac{FP_{ve}}{TP_{ve} + FP_{ve}}, Accuracy = 1 - \frac{FPR_{at} + FPR_{ve}}{2} \quad (5)$$

where the subscripts at and we refer to arterioles and venules, respectively. TP stands for true positive and FP stands for false positive. Since the first-stage evaluation just evaluates the performance of the vessel segmentation of the model, the results of the accuracy what follows in the paper refer to the $Accuracy$ in the second-stage if there is not special instruction.

To evaluate the effectiveness of the dual supervision, we investigate the performance of a model with supervision on the encoder module alone (Encoder-Supervised U-Net) and on the decoder module alone (Decoder-Supervised U-Net) by experiments, results can be seen in Table 2. Performance of Encoder-Supervised U-Net or Decoder-Supervised U-Net are not as good as dual DSN. Supervision on both the encoder module and the decoder module, dual supervision forms at this time, achieves the best results, and achieves an accuracy gain more than 2% compared with U-Net. From the visualization of the predicted image in Fig. 3, we can see that dual DSN predicts more fine-grained and more connectivity results.

Since the loss function in our proposed model is the sum of 9 loss functions, to improve the effect of the model, we conduct experiment that sets a small weight for the first loss function and raise the weight gradually with the deepening of the network. But this imbalanced weight will degrade the effect of the model. We will explore the adaptive weight for each loss function in future work.

Accuracy compared with others are shown in Table 3, our proposed model outperforms Xu et al. [11], which has the same evaluation strategy, with an accuracy gain of 5%, and also outperforms other methods which evaluated only on the centerline locations. Figure 4 shows an example of the classification results.

4   Conclusion

In this work, we present a novel deep-learning model for automated classification of arterioles and venules. We adapt U-shape network structure as our network backbone, and add dual supervision on the network. The experimental results show that the proposed dual DSN outperforms the previous state-of-the-art methods. The arteriovenous classification results produced by dual DSN are not only with more fine-grained details, but are also more generally realistic. In future work, we will explore the effect of dual supervision on other pixel-level classification tasks and further study the theoretical explanation of dual supervision.

References

1. AlBadawi, S., Fraz, M.M.: Arterioles and venules classification in retinal images using fully convolutional deep neural network. In: Campilho, A., Karray, F., ter Haar Romeny, B. (eds.) ICIAR 2018. LNCS, vol. 10882, pp. 659–668. Springer, Cham (2018). https://doi.org/10.1007/978-3-319-93000-8_75
2. Estrada, R., Allingham, M.J., Mettu, P.S., Cousins, S.W., Tomasi, C., Farsiu, S.: Retinal artery-vein classification via topology estimation. IEEE Trans. Med. Imaging 34(12), 2518–2534 (2015)
3. Hu, Q., Abràmoff, M.D., Garvin, M.K.: Automated separation of binary overlapping trees in low-contrast color retinal images. In: Mori, K., Sakuma, I., Sato, Y., Barillot, C., Navab, N. (eds.) MICCAI 2013. LNCS, vol. 8150, pp. 436–443. Springer, Heidelberg (2013). https://doi.org/10.1007/978-3-642-40763-5_54
4. Kondermann, C., Kondermann, D., Yan, M.: Blood vessel classification into arteries and veins in retinal images. In: Medical Imaging 2007: Image Processing, vol. 6512, p. 651247. International Society for Optics and Photonics (2007)
5. Ronneberger, O., Fischer, P., Brox, T.: U-Net: convolutional networks for biomedical image segmentation. In: Navab, N., Hornegger, J., Wells, W.M., Frangi, A.F. (eds.) MICCAI 2015. LNCS, vol. 9351, pp. 234–241. Springer, Cham (2015). https://doi.org/10.1007/978-3-319-24574-4_28

6. Shuangling, W., Yilong, Y., Guibao, C., Benzheng, W., Yuanjie, Z., Gongping, Y.: Hierarchical retinal blood vessel segmentation based on feature and ensemble learning. Neurocomputing **149**, 708–717 (2015)
7. Staal, J., Abràmoff, M.D., Niemeijer, M., Viergever, M.A., Van Ginneken, B.: Ridge-based vessel segmentation in color images of the retina. IEEE Trans. Med. Imaging **23**(4), 501–509 (2004)
8. Welikala, R., et al.: Automated arteriole and venule classification using deep learning for retinal images from the UK biobank cohort. Comput. Biol. Med. **90**, 23–32 (2017)
9. Xu, X., Ding, W., Abràmoff, M.D., Cao, R.: An improved arteriovenous classification method for the early diagnostics of various diseases in retinal image. Comput. Methods Programs Biomed. **141**, 3–9 (2017)
10. Xu, X., et al.: Smartphone-based accurate analysis of retinal vasculature towards point-of-care diagnostics. Sci. Rep. **6**, 34603 (2016)
11. Xu, X., et al.: Simultaneous arteriole and venule segmentation with domain-specific loss function on a new public database. Biomed. Opt. Express **9**(7), 3153–3166 (2018)
12. Zhang, J., Dashtbozorg, B., Bekkers, E., Pluim, J.P., Duits, R., ter Haar Romeny, B.M.: Robust retinal vessel segmentation via locally adaptive derivative frames in orientation scores. IEEE Trans. Med. Imaging **35**(12), 2631–2644 (2016)
13. Zhou, Z., Rahman Siddiquee, M.M., Tajbakhsh, N., Liang, J.: UNet++: a nested U-net architecture for medical image segmentation. In: Stoyanov, D., et al. (eds.) DLMIA/ML-CDS -2018. LNCS, vol. 11045, pp. 3–11. Springer, Cham (2018). https://doi.org/10.1007/978-3-030-00889-5_1

Liver Segmentation from Multimodal Images Using HED-Mask R-CNN

Supriti Mulay[1,2]([✉]) [iD], G. Deepika[3], S. Jeevakala[2], Keerthi Ram[2],
and Mohanasankar Sivaprakasam[1,2]

[1] Indian Institute of Technology Madras, Chennai, India
supriti@htic.iitm.ac.in
[2] Healthcare Technology Innovation Centre, Chennai, India
[3] Manipal Institute of Technology, Manipal, India

Abstract. Precise segmentation of the liver is critical for computer-aided diagnosis such as pre-evaluation of the liver for living donor-based transplantation surgery. This task is challenging due to the weak boundaries of organs, countless anatomical variations, and the complexity of the background. Computed tomography (CT) scanning and magnetic resonance imaging (MRI) images have different parameters and settings. Thus, images acquired from different modalities differ from one another making liver segmentation challenging task. We propose an efficient liver segmentation with the combination of holistically-nested edge detection (HED) and Mask- region-convolutional neural network (R-CNN) to address these challenges. The proposed HED-Mask R-CNN approach is based on effective identification of edge map from multimodal images. The proposed system firstly applies a preprocessing step of image enhancement to get the 'primal sketches' of the abdomen. Then the HED network is applied to enhanced CT and MRI modality images to get better edge map. Finally, the Mask R-CNN is used to segment the liver from edge map images. We used a dataset of 20 CT patients and 9 MR patient from the CHAOS challenge. The system is trained on CT and MRI images separately and then converted to 2D slices. We significantly improved the segmentation accuracy of CT and MRI images on a database with Dice value of 0.94 for CT, 0.89 for T2-weighted MRI and 0.91 for T1-weighted MRI.

Keywords: Liver segmentation · Holistically-nested edge detection · Mask-RCNN · Multimodal segmentation

1 Introduction

The liver is the largest digestive gland and detoxification organ in the human body. A CT and MRI are used to detect any injury or bleeding in the abdomen. This is a painless and accurate way to detect an internal trauma which helps in saving patients' lives. Automatic medical image segmentation approaches that are introduced in the last two decades have been the most successful methods for

© Springer Nature Switzerland AG 2020
Q. Li et al. (Eds.): MMMI 2019, LNCS 11977, pp. 68–75, 2020.
https://doi.org/10.1007/978-3-030-37969-8_9

medical image analysis. The feasibility of a CNN to be generalized to perform liver segmentation across various imaging strategies and modalities is used in [9]. Patrick et al. [2] presented a method to automatically segment liver and lesions in CT and MRI abdomen images using cascaded fully convolutional neural networks (CFCNs) enabling the segmentation of large-scale medical trials and quantitative image analysis. Liu et al. [5] proposed liver sequence CT image segmentation solution GIU-Net, which consolidates an improved U-Net and a graph cutting algorithm, to take care of the low contrast between a liver and its surrounding organs issue. The problem of the large difference among individual livers in CT image was also addressed in [5].

The principal approach to image segmentation is to detect image discontinuities, edges are one of those. Canny edge detection is the most popular technique for edge detection but has limitations that different scales not directly connected, also exhibit spatial shift and inconsistency [10].

HED was proposed by Xie et al. [10] to address these limitations. The original HED network was intended for edge discovery purposes in normal pictures, which catches fine and coarse geometrical structures (e.g. contours, spots, lines, and edges), while we are keen in capturing 'primal structure' in abdomen images.

We chose holistically-nested edge detection because it addresses the challenging ambiguity in edge and object boundary detection significantly. We proposed a unique method that can perform segmentation of liver on various modalities in detecting features and instance segmentation with a holistically nested edge (HED)-Mask R-CNN. We investigate a deep learning methodology that jointly detect the edges and then segments the liver. The network is trained on a subset of the CHAOS challenge and evaluated on other subset data of CHAOS challenge for both CT and MRI modalities.

Our contributions in the present work are,

- use of enhancement method to get 'primal sketches' of abdomen images
- utilize holistically-nested edge Mask RCNN (HED-Mask R-CNN) to get edge map
- applying the Mask R-CNN to segment liver from edge map images.
- lastly, we demonstrate the generalization and adaptability of HED-Mask R-CNN to different modalities

The remainder of the paper is described in following subsections. The Sect. 2 deals with joint network approach, Sect. 3 include the experiment and results, Sect. 4 presents a discussion of the proposed method and finally, Sect. 5 draws the conclusions of this work.

2   Joint Network Approach

The segmentation process of liver consists of joint deep learning pipeline: preprocessing, edge map detection (Fully convolutional network (FCN) with deep supervision) [10], feature extractor with fine-tuning layers [3], as depicted in Fig. 1.

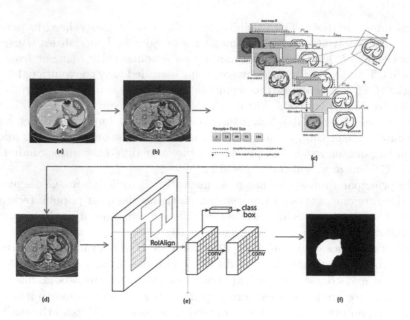

Fig. 1. Joint network architecture for automated semantic liver segmentation. (a) Original image (b) Enhanced image (c) HED network (d) The original image multiplied by the obtained edge map (e) Mask R-CNN. (f) Segmented liver output.

2.1   Pre-processing

Each DICOM slice is converted to PNG image and then pre-processing is carried out. Image noise, spatial resolution, and slice thickness affect CT and MRI images. An image enhancement technique is firstly applied, to get the organ sketches in abdomen images. We applied a separate enhancement technique to CT and MRI images because of their different resolutions. The CT images have been enhanced by modified sigmoid adaptive histogram equalization algorithm [7]. An adaptive histogram equalization (CLAHE) and sigmoid function are applied to preserve the mean brightness of the input CT images. We apply unsharp contrast enhancement filter to allow better differentiation of abnormal liver tissue in the case of MRI images. The abdomen organ features are enhanced prominently by this method. Fig. 2(a) and (c) shows original images of CT and MRI respectively and Fig. 2(b) and (d) shows the enhanced images.

2.2   Holistically-Nested Edge Detection Approach

Deep supervision used in HED that accounts for low-level predictions resulting in better edge map, is one of the reasons to consider HED in our method. We thus chose HED that automatically learns rich and important hierarchical representations from MRI/CT images to resolve the challenging ambiguity in edge and object boundary detection. It incorporates multi-scale and multi-level learning of deep

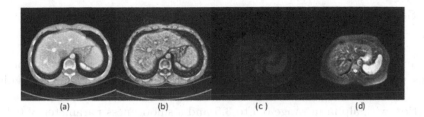

Fig. 2. (a) and (c) CT and MRI original images, (b) and (d) corresponding enhanced images

image features utilizing auxiliary cost functions at each convolutional layer. This network architecture is with 5 stages, including strides of 1, 2, 4, 8 and 16, can capture the inherent scales of organ contours [6]. Consequently, HED-based profound system models have been effectively utilized in medical image analysis for brain tumor segmentation [12], prostate segmentation [1], pancreas localization, and segmentation [6], retinal blood vessel segmentation [11].

The network structure is initialized based on an ImageNet pre-trained VGGNet model. The enhanced images are fed as an input to HED to get refined edge map. Organ edge/interior map predictions can be obtained at each side-output layer. The refined edge maps produced as side output are considered as an input to our next network. Superior output for each modality is chosen. A side output 6 is chosen for CT images whereas side output 0 for MRI images. Figure 3 shows the CT/MRI edge maps chosen to train the next CNN network for segmentation.

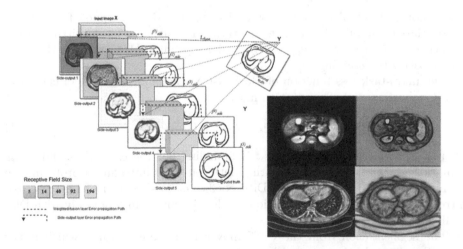

Fig. 3. CT and MR images with associated edge map images of training data

2.3 Segmentation with Mask R-CNN

2.3.1 Data Preparation

We use popular geometric augmentation techniques, flipping, and sharpening in this work. Elastic deformation distorts images locally by moving individual pixels around following a distortions field with strength sigma. We applied elastic distortion with alpha in range 0.5 to 3.5 and a smoothness parameter of 0.4.

2.3.2 Mask R-CNN

One of the most successful deep learning network for image segmentation is Mask R-CNN [3]. Therefore we chose this model to segment the liver from both modalities. Enhanced image multiplied with obtained edge map from HED is used as input for Mask R-CNN to understand the 3D structure. Prior knowledge of edge map gives advantage in segmenting objects with a large variety in appearance and lack of texture to strong textures. We have used an end-to-end pre-trained Mask R-CNN model with a Resnet-101-FPN backbone in this study. This model has been pre-trained on Imagenet dataset. It predicts the masks of detected regions and classifies them into one of the classes given at the time of training. We choose an existing open-source implementation [8] using Tensorflow deep learning framework.

The inputs for HED FCN are gray-scale images of size 512×512 and their outputs are images of size $512 \times 512 \times 3$. The model is implemented in Keras[1] with the TensorFlow[2] backend.

2.3.3 Training Strategy

Even though Mask R-CNN is profound enough and is equipped for learning appropriate parameters for liver segmentation, it is inclined to over-fitting issues. We utilize an effective technique such as Adam optimizer [4] to alleviate this issue and boosting the training.

The multi-task loss function of Mask R-CNN combines the loss of classification, localization and segmentation mask.

$$L = L_{cls} + L_{box} + L_{mask} \tag{1}$$

where L_{cls} is classification loss, L_{box} is bounding box regression loss and L_{mask} is mask loss. Dice coefficient performs better at class imbalanced problems. So we modified the L_{mask} loss with Dice coefficient loss instead of binary cross-entropy loss. We observed that validation loss is converging smoothly with Dice coefficient loss.

The segmentation from the CNN may contain some artifact which is not liver. To relieve this issue, some basic post-preparing was performed.

---

[1] https://keras.io/.
[2] https://www.tensorflow.org/.

3   Experiment and Results

Results of the automated liver segmentation are exhibited in Fig. 4. Comparison of ground truth with the segmented liver is highly promising for obtaining high-performance metrics. The whole setup was implemented in Linux environment using NVIDIA GTX 1080 8 GB GPU on a system with 16 GB RAM and having Intel Core-i5 7th generation @3.20 GHz processor.

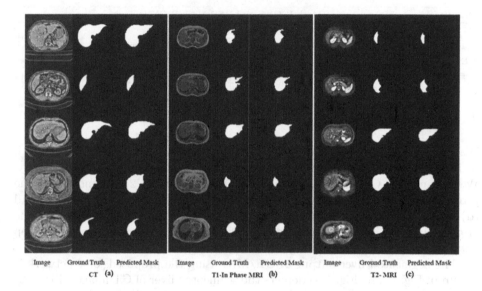

Image   Ground Truth   Predicted Mask   Image   Ground Truth   Predicted Mask   Image   Ground Truth   Predicted Mask

CT   (a)   T1-In Phase MRI   (b)   T2- MRI   (c)

Fig. 4. CT and MR images with their respective ground truth and predicted output.

3.1   Datasets

The network training is run on a subset of the publicly available CHAOS challenge[3] containing data sets from two different modalities. The training is applied to 2183 axial CT slices (16 patients), 452 (4 patients) are used for validation. For MRI T2-weighted we applied training on 469 axial (7 patients) images and 80 (2 patients) images are used for validation. In the case of T1-weighted in-phase images, we use 316 axial (15 patients) images for training and 105 (5 patients) images for testing.

3.2   Evaluation

For each image, detections made by the model are compared to the ground truth label to evaluate the model performance given in the challenge for that image. We want to demonstrate the robustness, generalization, and scalability of our

---

[3] https://chaos.grand-challenge.org/.

proposed method in this work. Table 1 provides the comparison between Mask R-CNN and HED-Mask R-CNN. Comparison with ground truth and segmented liver give rise to the assertion that our approach is highly promising for obtaining high-performance metrics. Dice coefficient of 0.9 above for both CT and MRI images shows that our method works well.

Table 1. Comparison of liver segmentation on CT and MRI scans

Quantitative comparison among Mask R-CNN and HED-Mask R-CNN			
CT Segmentation	Dice	MRI Segmentation	Dice
Mask R-CNN(N=491)	0.90	Mask R-CNN (N = 105 T1-weighted)	0.80
HED-Mask R-CNN(N = 491)	0.94 ± 0.03	HED-Mask R-CNN (N = 105 T1-weighted)	0.91 ± 0.06

N = number of slices

4   Discussion

We propose a combination of HED (deep version) and Mask R-CNN network to improve the liver segmentation performance of CT/MRI imaging modality. We enhance the CT/MRI images which are shown in Fig. 2 to demonstrate the proficiency of the proposed joint deep network. It is seen from the Fig. 2(b) and (d) that the edge information and contrast of the liver are enhanced than the original images. The enhanced images of CT/MRI are fed to the HED deep network to extract the edge map as shown in Fig. 3.

The segmented liver output of CT/MRI using the proposed joint network is shown in Fig. 4. The Fig. 4(a) depicts the segmented liver of CT images, Fig. 4(b) depicts the segmented liver of T1-in phase and Fig. 4(c) depicts the segmented liver of T2 MRI images. The segmentation results of our proposed joint deep network are compared with the ground truth extracted by the medical experts. It is observed that our proposed networks perform well in segmentation. The comparison of Dice value for Mask R-CNN and proposed HED-Mask R-CNN is tabulated in Table 1. It is observed from Table 1 the CT images without HED network obtained the Dice coefficient of 0.90 and with HED the Dice coefficient of 0.94. Similarly, for MRI images the Dice coefficient without and with HED network are 0.80 and 0.91 respectively. It is seen from Table 1 that HED network with the combination of Mask R-CNN increases the segmentation accuracy of the proposed network.

The post-processing using graph cut method requires initial segmentation image with larger liver area and controlling parameters needs to be determined by multiple experiments in GIU-Net [5]. Whereas we present a framework, which is capable of a segmenting the liver with no post-processing strategy. Also generalized CNN [9] relied on retrospective data to train and validate the multimodal CNN, in contrast our method takes the current data to train the model. Thus, the proposed joint deep networks outperform well for both CT and MRI modalities of images.

5   Conclusions

We demonstrate our strategy for liver segmentation in multimodal CT and MRI images using HED-Mask R-CNN method in this work. The novelty of our work is in the use of edge map with Mask R-CNN with automatic features learning instead of just applying CNN for object segmentation. Importantly, the above framework obviates the need for liver segmentation significantly increasing robustness and accuracy as compared to stand-alone segmentation methods. Our method yields Dice 0.94 for CT and 0.91 for MRI images. Our results on 491 CT slices and 105 slices demonstrate an impressive improvement over independent CNN based strategies and may give significant clinical estimations for liver segmentation. We intend to apply our technique to segment liver from additional imaging modalities and all organs segmentation as well.

References

1. Cheng, R., et al.: Automatic magnetic resonance prostate segmentation by deep learning with holistically nested networks. J. Med. Imaging (Belling-ham, Wash.) **4**(4) (2017). 041302
2. Christ, P.F., et al.: Automatic liver and tumor segmentation of CT and MRI volumes using cascaded fully convolutional neural networks. arXiv preprint arXiv:1702.05970 (2017)
3. He, K., Gkioxari, G., Dollár, P., Girshick, R.: Mask R-CNN. In: 2017 IEEE International Conference on Computer Vision (ICCV), pp. 2980–2988 (2017)
4. Kingma, D., Ba, J.: Adam: a method for stochastic optimization. In: Proceedings of the 3rd International Conference on Learning Representations (ICLR) (2015)
5. Liu, Z., et al.: Liver CT sequence segmentation based with improved U-Net and graph cut. Expert Syst. Appl. **126**, 54–63 (2019). https://doi.org/10.1016/j.eswa.2019.01.055
6. Roth, H.R., et al.: Spatial aggregation of holistically-nested convolutional neural networks for automated pancreas localization and segmentation. Med. Image Anal. **45**, 94–107 (2017). https://doi.org/10.1016/j.media.2018.01.006
7. Mulay, S., Ram, K., Sivaprakasam, M., Vinekar, A.: Early detection of retinopathy of prematurity stage using deep learning approach. In: Proceedings of SPIE 10950, Medical Imaging 2019: Computer-Aided Diagnosis, p. 109502Z (2019). https://doi.org/10.1117/12.2512719
8. Abdulla, W.: Mask r-CNN for object detection and instance segmentation on Keras and TensorFlow (2017). https://github.com/matterport/Mask_RCNN
9. Wang, K., et al.: Automated CT and MRI liver segmentation and biometry using a generalized convolutional neural network. Radiol.Artif. Intell. **1**(2), 180022 (2019). https://doi.org/10.1148/ryai.2019180022
10. Xie, S., Tu, Z.: Holistically-nested edge detection (2015). https://doi.org/10.1109/ICCV.2015.164
11. Lin, Y., Zhang, H., Hu, G.: Automatic retinal vessel segmentation via deeply supervised and smoothly regularized network. IEEE Access **7**, 57717–57724 (2019). https://doi.org/10.1109/ACCESS.2018.2844861
12. Zhuge, Y., et al.: Brain tumor segmentation using holistically nested neural networks in MRI images. Med. Phys. **44**(10), 5234–5243 (2017). https://doi.org/10.1002/mp.12481

aEEG Signal Analysis with Ensemble Learning for Newborn Seizure Detection

Yini Pan[1], Hongfeng Li[2,4], Lili Liu[3], Quanzheng Li[5], Xinlin Hou[3(✉)],
and Bin Dong[4,6(✉)]

[1] Academy for Advanced Interdisciplinary Studies, Peking University,
Beijing 100871, China
[2] Center for Data Science in Health and Medicine, Peking University, Beijing, China
[3] Department of Pediatrics, Peking University First Hospital, Beijing 100034, China
houxinlin66@163.com
[4] Beijing Institute of Big Data Research, Beijing, China
[5] Gordon Center for Medical Imaging, Massachusetts General Hospital
and Harvard Medical School, Boston, MA 02114, USA
[6] Beijing International Center for Mathematical Research, Peking University,
Beijing 100871, China
dongbin@math.pku.edu.cn

Abstract. Amplitude-integrated EEG (aEEG) has been widely used in neonatal seizure monitoring due to its convenience and broad applicability. However, due to the long length of aEEG signals, detecting seizures in aEEG signals is still a challenging and time-consuming task for experienced clinicians. In this paper, we propose an ensemble learning algorithm to tackle with this problem, aiming to assist clinicians to identify seizures more efficiently and effectively. Firstly, we employ wavelet denoising method to improve the signal-noise rate (SNR) of aEEG signal. Then, to reduce the high dimensionality of aEEG signals while retaining the essential information, we extract global and local features from aEEG signals based on visual features and entropy. Thereafter, we process our data with a feature augmentation algorithm to obtain an extended data set. Finally, an ensemble algorithm is utilized to perform seizure detection. We conduct experiments on real clinical data collected from Peking University First Hospital. Experimental results show that the proposed algorithm achieves excellent performance in seizure detection.

Keywords: Seizure detection · aEEG signal · Ensemble learning

1 Introduction

Neonatal seizure is one of the most critical symptoms of neonatal nervous system and one of the most common clinical manifestations of neonatal neurological abnormalities. The immature brain tissue of the newborn is susceptible

Y. Pan, H. Li and L. Liu—Equally contributed.

to injury, and frequent seizures may cause convulsive brain damage. Therefore, timely detection of neonatal seizures and treatment can reduce the occurrence of convulsive brain injury and improve the neurodevelopmental prognosis.

Electroencephalogram (EEG) is the gold standard for the diagnosis of neonatal seizures [11]. Traditional EEG requires that large number of electrodes be placed on the patients' scalp, which is difficult to perform for newborns. In addition, due to the large number of electrodes and lots of information collected, it is easy to adulterate some artifacts while collecting EEG signals. Therefore, the interpretation of conventional EEG requires a large amount of training.

The amplitude of EEG changes during convulsions, and can appear as a transient increase in the upper and lower boundaries on aEEG. In clinical work, it is recommended that neonatal doctors mark the aEEG fragments they believe to be suspicious as "onset" for confirmation in the corresponding original EEG. A number of studies showed that aEEG has comparable sensitivity and specificity with EEG but is much more practical in diagnosis [8]. Thus, we focus on aEEG signals for seizure detection.

Machine learning has been widely studied and applied in medical signal processing. There have been multiple studies on the topic of EEG classification [3]. However, aEEG classification has just developed recently and only a few methods have been proposed for seizure detection [16]. Although existing machine learning methods have achieved high accuracies in some tasks, these methods mainly rely on human interactions in the data preprocessing stage which limits their implementation in actual clinical workflow.

Classifying aEEG signals under a clinically implementable scenario remains a great challenge. Firstly, aEEG signals are often much longer than EEG signals. Labeling every seizure onset in an aEEG signal is time-consuming due to its long length. Therefore, we have to deal with very long signals with only global (or weak) labels. Secondly, different from EEG signals, aEEG signals do not have fixed length. As a result, automatically extracting features from aEEG signals with deep learning methods is difficult. Thirdly, aEEG signals have only one channel, which means that they can be easier disturbed by the environment than the EEG signals.

To resolve the above issues, in this paper, we propose a novel method to extract features from aEEG signals and employ an ensemble algorithm to classify aEEG signals. Our algorithm is able to deal with aEEE signals with variable lengths and small size. Experimental results show that the proposed algorithm can achieve a promising performance in the task of aEEG signal classification for seizure detection. The highlights of this paper are listed as follows:

(1) We propose using the fuzzy entropy to select the top and negative instances to effectively represent an aEEG signal and introduce feature augmentation to expand data set.

(2) The proposed algorithm, i.e. feature selections followed by ensemble learning, achieves excellent classification performance and provides a good balance between the specificity and sensitivity.

The remainder of the paper is organized as follows. Section 2 introduces the collected data. The seizure classification algorithm is described in detail in Sect. 3. Evaluations of the proposed algorithm on the real clinical aEEG data are presented in Sect. 4. Finally, we draw conclusions in Sect. 5.

2   Data Acquisition and Introduction

The aEEG data used in this paper were collected from the neonatal ward of Peking University First Hospital, with 216 newborns with neonatal seizures and 310 newborns with normal aEEG results. aEEG was undertaken by bedside of the newborn and trained nurses in the neonatal ward were responsible for aEEG recording operation and observing clinical seizures during the examination. The aEEG reports were issued by neonatal physician with aEEG experience over 5 years, who were not informed about the condition of the patients and were not involved in the clinical diagnosis and treatment in order to ensure the objectivity of the reports. Figure 1 demonstrates the characteristics of aEEG recorded.

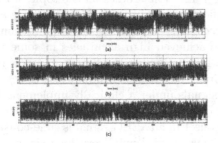

Fig. 1. Characteristics of aEEG. (a) Normal background pattern; (b) Discontinuous background pattern; (c) Burst suppression background pattern.

3   Seizure Detection Algorithm

In this section, we describe the proposed algorithm in detail. The flowchart of the algorithm is shown in Fig. 2.

Fig. 2. The flowchart of the proposed algorithm.

3.1   Feature Selection

Feature extraction plays an important role in machine learning tasks. Here, we extract two types of features from the aEEG signals, i.e., global features and local features. The global features include 4 basic features and total histogram. Basic features include minimum, maximum, skewness and kurtosis of the whole aEEG signal. Total histogram is calculated with the algorithm in [6]. As for the local features, a sliding window with the length of 3 min is applied to the aEEG signal to capture sudden changes. The overlap between two successive windows is set to 1.5 min. Then, local features are calculate in each window.

Here, we employ several entropies to extract the features of an aEEG signal. These include auto permutation entropy (APE) [13], sample entropy [2], approximate entropy [12], fuzzy entropy [15]. Besides, we introduce spectral entropy to add spectral information into our feature. We also compute the lower and upper border of an aEEG window. The lower border and upper border are defined as the mean of five points near the maximum and minimum values of an aEEG window's envelope, respectively. Together with the entropy features, the desired local features of an aEEG signal are obtained. Furthermore, with the window selection that will be described later in this paper, we get 22 windows for every signal. Thus, there are 236 columns for each local feature in total.

3.2   Top Instance and Negative Evidence for Windows Selection

Generally speaking, seizures can happen at any time in an aEEG signal, and the length of the signal can be very long. Thus, detecting seizures in an aEEG by going through every second of the whole signal is nearly impossible. However, cutting off the entire signal into shorter ones with the same length (for instance, 3 h) may degrade the performance of classifiers.

Here we perform the following operations: (1) if an aEEG signal exceeds 3 h, then we assume that there must be at least one seizure during the last 3 h. So we truncate the aEEG signal and keep the last 3 h. (2) if an aEEG signal is less than 3 h, then we pad the upper and lower boundaries of aEEG by the corresponding upper and lower boundaries' mean values with random number in the range of [0, 1]. After truncating and padding, we observe that our AUC score improves by 3% compared to the original aEEG series.

Inspired by the top and negative sampling [7], we introduce a new window selection algorithm to tackle the problem caused by the variable aEEG lengths in seizure detection tasks.

We further demonstrate the top and negative windows of aEEG signals mentioned above in Fig. 3. The upper curve is the original aEEG data and the lower curve is the aEEG signal with selected windows. We use the red curve to denote the top seizure-like windows chosen by our algorithm, the green curve to denote the negative windows and the red arrow to denote the seizure onset pointed by doctors. As the figure shows, top windows appear in those places where there are abrupt changes, which exactly corresponds to the most probable seizure onsets.

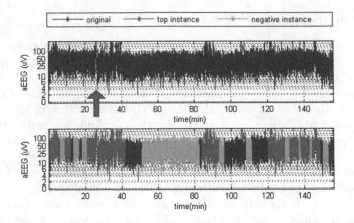

Fig. 3. aEEG signals with selected windows. The upper curve is the original signal and the lower curve is the signal after window selection. Red area indicates the top windows, green area indicates the negative evidences, and the red arrow indicates where seizure happens confirmed by clinicians. (Color figure online)

We calculate certain entropies of the signal within each window and sort the windows according to their corresponding entropy values. Then, only the first k windows and last k windows are retained for further feature extraction. After many trails, we find optimal $k = 11$ and the best entropy is fuzzy entropy. Classification results with different window selection entropy are shown in Table. 1.

Table 1. Classification results with different window selection entropy

Method	Accuracy	Specificity	Sensitivity	AUC
Approximate entropy	77.38%	77.91%	75.27%	75.27%
Sample entropy	77.24%	77.51%	75.10%	75.10%
Shannon entropy	76.83%	76.95%	74.63%	74.41%
Fuzzy entropy	79.73%	79.96%	79.01%	78.34%
Permutation entropy	78.25%	78.55%	76.29%	76.29%

3.3 Feature Augmentation

Data augmentation is a powerful method in supervised learning. Here, we apply interpolation and extrapolation to augment our train set and measure on our test set as: (1) Interpolation: for each data point x, find k nearest neighbors with the same label and interpolate by $x_{new} = (x - x_k) * \lambda + x_k$; (2) Extrapolation: for each data point x, find k nearest neighbors with the same label and extrapolate by $x_{new} = (x_k - x) * \lambda + x$. Here, k and λ are trainable parameters and we set $k = 9$ and $\lambda = 0.73$. Experimental results show that extrapolation performs better than interpolation and improves the AUC score by 0.6%.

3.4   Classification Algorithms

When merging features, we apply min-max normalization to data to improve the speed and accuracy. Then, with the extracted features, we adopt several algorithms, i.e., support vector machine (SVM) [1], logistic regression (LR) [10], adaptive boosting (AdaBoost) [14], Xgboost [5], random forest (RF) [4] and gradient boosting trees (GBDT) [9], as classifiers to classify the aEEG signals.

3.5   Ensemble Algorithm

Ensemble learning algorithm combining multiple machine learning algorithms together commonly leads to better performance than any of the individual classifier. Therefore, we find the best single classifier and use bagging strategy to form an ensemble learning algorithm for aEEG signal classification. Experimental results demonstrate that the ensemble algorithm can indeed further improve the classification accuracy.

4   Experiments and Analysis

In this section, we conduct experiments on the real data set described in Sect. 2 using the algorithm proposed in the previous section.

Prior to feature extraction, the aEEG signals are denoised with a band-pass filter to cut-off frequencies lower than 0.3 Hz and higher than 30 Hz to remove artifacts and a four-order wavelet is utilized to further improve the quality of the signals. In the experiments, we randomly divide the entire data set into three subsets, i.e., training, validation and testing sets. Specifically, we first randomly divide the entire data set into two parts with proportions of 80% and 20%, respectively. The 20% portion (105 samples) is used for testing. Then, 50 samples is further taken from the 80% portion for validation. Thus, the remaining 421 samples are used for training. We repeat the procedure 15 times and calculate the mean classification accuracy. All the experiments are carried out on a Dell laptop with Python 3.6 and MATLAB 7.0.

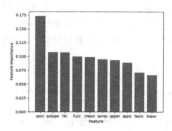

Fig. 4. Feature importance

4.1 Feature Evaluation

Figure 4 demonstrates the importance of features. We observe that the spectral entropy is of the most importance. Features like basic feature, lower border and approximate entropy are not as important as expected. As for the fuzzy entropy, it may be more significant in local context for window chosen in a single signal rather than sample comparison. The reason is that different patients may have a very different background signal level but possess a similar fuzzy entropy score.

4.2 Comparison of Classification Algorithms

Commonly, in medical image processing, positive means ill and negative means normal. We use following metrics for model evaluation and comparison: (1) $Accuracy = \frac{TP+TN}{TP+TN+FP+FN} * 100$; (2) $Sensitivity = \frac{TP}{TP+FN} * 100$; (3) $Specificity = \frac{TN}{TN+FP} * 100$; (4) Receiver operating characteristic curve (ROC). Here, TP indicates true positive, TN indicates true negative, FP indicates false positive and FN indicates false negative.

Table 2. Classification results with various classifiers.

Model	Accuracy	Specificity	Sensitivity	AUC
SVM	73.28%	73.46%	71.01%	71.01%
LR	74.08%	76.45%	78.26%	60.68%
Adaboost	76.79%	76.77%	75.46%	75.46%
Xgboost	78.12%	78.77%	78.70%	75.91%
Random Forest	74.53%	76.45%	76.54%	74.57%
GBDT	79.23%	79.27%	77.67%	77.67%
Ensemble	79.73%	79.96%	79.01%	78.34%

With the extracted features for aEEG signals, we compare the performance achieved with the six classification algorithms. Results are shown in Table 2. From Table 2 we conclude that the proposed feature extraction algorithm is effective for extracting semantic features from aEEG signals. Note that the ensemble algorithm made up of bagging of GBDTs obtains the best results in all categories. Table 3 shows that our work is better than the past researches and improves the accuracy by about 3%.

Table 3. Classification results compared with other researches.

Model	Accuracy	Specificity	Sensitivity	AUC
Wang [6]	76.31%	76.59%	74.59%	74.91%
Yang [16]	76.23%	76.30%	74.86%	74.86%
Our work	79.73%	79.96%	79.01%	78.34%

5   Conclusions

aEEG can be effective for screening newborns with high risk factors of neonatal seizures. In this paper, we propose a novel algorithm to extract features from aEEG signals and performed seizure classification using an ensemble method. Firstly, we utilize a method based on visual features and entropy to extract the global and local features from aEEG signals. Then, we expand our data with feature augmentation method. Finally, a bagging of GBDT models is employed to perform seizure detection. Experimental results on a real aEEG data set show that the ensemble algorithm can achieve a promising performance with the classification accuracy of 79.73%, the specificity score of 79.96%, the sensitivity score of 79.96% and the AUC score of 78.34%.

Acknowledgement. This work was supported in part by the National Key Research and Development Program of China under Grant 2018YFC0910700 and in part by the National Natural Science Foundation of China under Grants 11701018, 11831002 and 81801778.

References

1. Adankon, M.M., Cheriet, M.: Support vector machine. Comput. Sci. **1**(4), 1–28 (2002)
2. Bai, D., Qiu, T., Li, X.: The sample entropy and its application in eeg based epilepsy detection. J. Biomed. Eng. **24**(1), 200 (2007)
3. Beyli, E.D.: Combined neural network model employing wavelet coefficients for EEG signals classification. Digital Signal Process. **19**, 297–308 (2009)
4. Breiman, L.: Random forest. Mach. Learn. **45**, 5–32 (2001)
5. Chen, T., He, T., Benesty, M., et al.: XGBoost: extreme gradient boosting. Rpackage version 0.4-2 pp. 1–4 (2015)
6. Chen, W., Wang, Y., Cao, G., Chen, G., Gu, Q.: A random forest model based classification scheme for neonatal amplitude-integrated EEG. Biomed. Eng. Online **13**(2), S4 (2014)
7. Courtiol, P., Tramel, E.W., Sanselme, M., Wainrib, G.: Classification and disease localization in histopathology using only global labels: a weakly-supervised approach. arXiv preprint arXiv:1802.02212 (2018)
8. Frenkel, N., et al.: Neonatal seizure recognition-comparative study of continuous-amplitude integrated EEG versus short conventional EEG recordings. Clin. Neurophysiol. **122**(6), 1091–1097 (2011)
9. Friedman, J.H.: Greedy function approximation: a gradient boosting machine. Ann. Stat. **29**(5), 1189–1232 (2001)
10. Menard, S.: Applied logistic regression analysis. Technometrics **38**(2), 192 (2002)
11. Nagarajan, L., Palumbo, L., Ghosh, S.: Classification of clinical semiology in epileptic seizures in neonates. Eur. J. Paediatr. Neurol. **16**(2), 118–125 (2012)
12. Pincus, S.: Approximate entropy (ApEn) as a complexity measure. Chaos Interdisc. J. Nonlinear Sci. **5**(1), 110 (1995)
13. Ping, X., Xiuli, W., Yihao, D.: Feature extraction method of semg based on auto permutation entropy. PR&AI **27**(6), 496–501 (2014)

14. Schapire, R.E.: The boosting approach to machine learning: an overview. In: Denison, D.D., Hansen, M.H., Holmes, C.C., Mallick, B., Yu, B. (eds.) Nonlinear Estimation and Classification, pp. 149–171. Springer, New York (2003). https://doi.org/10.1007/978-0-387-21579-2_9
15. Xiang, J., et al.: The detection of epileptic seizure signals based on fuzzy entropy. J. Neurosci. Methods **243**, 18–25 (2015)
16. Yang, T., Chen, W., Cao, G.: Automated classification of neonatal amplitude-integrated EEG based on gradient boosting method. Biomed. Signal Process. Control **28**, 50–57 (2016)

Speckle Noise Removal in Ultrasound Images Using a Deep Convolutional Neural Network and a Specially Designed Loss Function

Danlei Feng[1,3], Weichen Wu[2,3], Hongfeng Li[4], and Quanzheng Li[5(✉)]

[1] Academy for Advanced Indisciplinary Studies, Peking University, Beijing, China
[2] Yuanpei College, Peking University, Beijing, China
[3] Beijing Institute for Big Data Research, Beijing, China
[4] Center for Data Science in Health and Medicine, Peking University, Beijing, China
[5] Gordon Center for Medical Imaging, Massachusetts General Hospital
and Harvard Medical School, Boston, MA 02114, USA
quanzhengli5@gmail.com

Abstract. The removal of speckle noise in ultrasound images has been the focus of a number of researches. Meanwhile, deep convolutional neural networks (DCNN) has been proved effective for various computer vision tasks, including image classification, segmentation and denoising. In this paper, we apply deep convolutional neural network to remove speckle noise in ultrasound images. Besides, a new hybrid loss function is specially designed for speckle noise removal, which can result in faster and more stable convergence during training. Experiments on synthetic and real Ultrasound images show that the proposed model outperforms other speckle reduction methods.

Keywords: Ultrasound images · Speckle noise removal · Deep neural networks

1 Introduction

In recent years, Ultrasound diagnostic technology has been widely used to assist in the diagnosis of diseases in the abdomen, thyroid, breast, uterus and other places due to its safety, low cost and real-time imaging. However, Ultrasound images are corrupted with inherent speckle noises that show a granular appearance, which impairs the information of images and makes it unclear for human to distinguish the vital tissue or image features in diagnosis [9]. Besides, speckle noise can also enhance the difficulty of other ultrasound image processing tasks such as classification [3] and segmentation [1]. Therefore, speckle noise removal

This work was supported in part by the National Key Research and Development Program of China under Grant 2018YFC0910700 and in part by the National Natural Science Foundation of China under Grants 11701018, 11831002 and 81801778.

© Springer Nature Switzerland AG 2020
Q. Li et al. (Eds.): MMMI 2019, LNCS 11977, pp. 85–92, 2020.
https://doi.org/10.1007/978-3-030-37969-8_11

becomes an important step for analyzing and processing medical ultrasound images.

Many researchers focused on the mathematical model of speckle noise in ultrasound images. It is widely recognized that the distribution of speckle noise is signal related. [14] suggests that (1) is the most suitable for ultrasound speckle noise.

$$\tilde{x} = x + \sqrt{x} * n, \tag{1}$$

Here, $x \in \mathbb{S} = (0, M]^{m \times n}$ is the original image, $\tilde{x}$ is the noisy image, and n is Gaussian noise $N(0, \sigma^2)$.

Various methods have been proposed for the removal of ultrasound speckle noise such as local statistics [6,11,12], anisotropic diffusion methods [10,20], nonlocal mean approaches [2,4] and variational approaches [7,17]. Lee filter [12], Frost filter [6], and Kuan filter [11] are the most classic methods to despeckle images. But they are more likely to get smooth images without preserving the edge and other features. Variational approaches [7,17] are one of the most popular branches in ultrasound despeckling. These methods transfer the denoising task into a convex optimization problem:

$$r_{CVX}(\tilde{x}) = \underset{x}{\mathrm{argmin}}[\lambda D(x, \tilde{x}) + P(x)], \tag{2}$$

where $D(x, \tilde{x})$ is the data fidelity term that measures the similarity between the denoised image and the noisy image. $P(x)$ is the prior term that measures the reality of the denoised image and λ controls their relative importance. Iterative optimization algorithms, most typically alternative direction method of multipliers (ADMM), are then applied to solve the optimization problem. Non local mean methods [2] are also popularly used because they can remove speckle noise while preserving the texture details. Block-matching 3D Filtering (BM3D) [5] is the most widely used approach for Gaussian denoising.

In recent years, deep learning approaches have been introduced for image denoising, such as DnCNN [21], WIN [13], ELU-CNN [18], Cascaded CNN [19]. Different deep learning methods vary in their network architectures and choices of loss functions, which can make different achievements. Neural network approaches can achieve the state-of-the-art results in denoising for natural images, CT images and MRI images, all with Gaussian noise or Gaussian mixed noise. In this work, we introduce a new convolutional neural network and a new hybrid loss function for ultrasound noise removal, which outperforms other ultrasound denoising methods.

The remainder of this paper is organized as follows. Section 2 describes the proposed method for ultrasound image denoising; Sect. 3 contains experimental results on both synthetic and real ultrasound images. Section 4 concludes the our work.

2   The Proposed Method

In this section, we first illustrate the architecture of the proposed US-Net (Neural network for Ultrasound) model that is designed especially for Ultrasound removal. Then, we propose a new hybrid loss and show that the new loss suits well for ultrasound despeckling task.

2.1 Model

Inspired by the CNN-based model that successfully improved the results of Gaussian noise removal, we propose a new model named US-Net to solve the Ultrasound denoising problem. The network structure of US-Net is shown in Fig. 1.

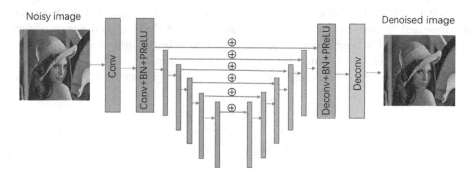

Fig. 1. Proposed USNet

Our model includes 6 convolutional layer blocks, 6 symmetric deconvolutional layer blocks and 2 single convolutional and deconvolutional layers. The convolutional block composed of convolution, batch normalization [8] and ReLU layers. And the symmetric deconvolution block consists of deconvolution, BN and ReLU layers. Convolution blocks are used for feature extraction. Deconvolution was first used as the neural network layer in the field of semantic segmentation [16]. It can recover more details of image contents as the replacement of upsampling layer, which repeats the adjacent pixel blocks and as tends to smooth the images. Skip connections are added from a convolutional block to its symmetric deconvolutional block which is schematically represented in Fig. 1. By horizontal connections. In this way we can add more fine-grained details that may be lost during forward propagation to the deconvolutional layer and thus get fine outputs with less information loss.

2.2 Loss Function

Mean square error (MSE), defined as (3), is the most frequently used loss function for CNN-based image denoising models [19, 21–23] due to its

$$L_{MSE} = \frac{1}{2N} \sum_{i=1}^{N} \|F(\tilde{x}_i) - x_i\|^2, \tag{3}$$

convex and differentiable properties that are significant for optimization problems. But it is widely noticed that MSE is not always consistent with human visual system (HVS) [24] and the formula of MSE loss works under the assumption of white Gaussian noise instead of speckle noise. Thus we propose a new hybrid loss which is defined as

$$\mathscr{L}_{train} = \alpha \mathscr{L}_{US} + (1 - \alpha) \mathscr{L}_{SSIM} \tag{4}$$

Here, $\mathscr{L}_{US}$ is the new loss designed for ultrasound speckle noise and $\mathscr{L}_{SSIM}(\tilde{x}_i, x_i) = \sum(1 - SSIM(F(\tilde{x}_i), x_i))$ is the error summation of SSIM [24] with respect to the denoised image and original image. These two parts are designed to control the contents and structure of the denoised image respectively, and α is the weight value to balance their relative importance.

$\mathscr{L}_{US}$ is especially designed for speckle noise removal.

According to the ultrasound speckle noise model (1), we have:

$$
\begin{aligned}
\sqrt{\tilde{x}_i} &= \sqrt{x_i + \sqrt{x_i} n_i} \\
&= \sqrt{x_i} \sqrt{(1 + \frac{n_i}{x_i})} \\
&= \sqrt{x_i}(1 + \frac{n_i u_i}{2} + o(n_i u_i)) \\
&= \sqrt{x_i} + \frac{n_i}{2} + o(n_i) \\
&= \sqrt{x_i} + \frac{n_i}{2} + o(\sigma)
\end{aligned}
\tag{5}
$$

This inspires us to use

$$\mathscr{L}_{US}(\tilde{x}_i, x_i) = \frac{1}{2N} \sum_{i=1}^{N} \| \sqrt{F(\tilde{x}_i)} - \sqrt{x_i} \|^2 \tag{6}$$

where F denotes the denoising neural network function which maps the noisy image $\tilde{x}_i$ to the denoised image $F(\tilde{x}_i)$.

Besides, to test the convergence and stability, we ran an comparative experiment in which the same network is trained with $\mathscr{L}_{train}$ and MSE loss respectively. Figure 2 shows the training val loss of different losses. The network trained with $\mathscr{L}_{train}$ loss converges in a more stable and faster way than the one trained with MSE.

2.3   Training Setup

We implement our proposed network with Adam optimization. The convolution kernels was 3×3 with a feature map of 64 channels. Zero padding was used to keep the same size. The learning rate starts from 10^{-3} and reduces to 10^{-4} after 30 epoch. The mini-batch size is set as 8. The network were trained in Keras based on Tensorflow backend on a GTX1080Ti GPU.

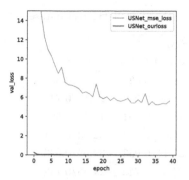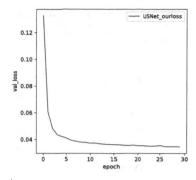

Fig. 2. Val loss values of $\mathscr{L}_{train}$ and MSE loss on the same network

3   Experimental Results

In this section, we present the results of our proposed US-Net model on both synthetic and real Ultrasound images. In order to demonstrate the effectiveness of the proposed method for speckle noise removal, we compare it with the following seven approaches: Lee filter [12], Frost filter [6], Kuan filter [11], anisotropic diffusion method [20], non local mean filter [2], BM3D [5] and DnCNN [21]. DnCNN is the only one except our method using neural network in our comparisons, which achieves state-of-the-art performance for Gaussian denoising. The peak signal to noise ratio (PSNR) and structural similarity index (SSIM) [?] are calculated to evaluate the denoising performance of different methods.

3.1   Results on Synthetic Images

Followed by [21], we used the Berkeley segmentation dataset (BSD500) [15] for training and Set12 dataset for testing. Data augmentation (rotation or flip) and random cropping were used for all the images so that their size is 180 × 180 pixel. Different variance of Gaussian noise in (1) were added for comparison. The numerical results are listed in Table 1. It can be seen that the proposed US-Net yields the highest PSNR and SSIM on different noise levels. The visual results of the denoised images are shown in Fig. 3. We can observe that our proposed model keeps sharp edges and fine details and outperforms other approaches.

3.2   Results on Real Ultrasound Images

We used the thyroid Ultrasound dataset from Peking University Hospital for training and testing, during which the images were cropped into 9124 patches with 180 × 180 sizes. Experiments are carried out on two different noise variances. The numerical results are shown in Table 2. We can notice that the PSNR and SSIM of our proposed method also get the best results. And visual results are illustrated in Fig. 4.

Table 1. Numerical results of different methods on Set12 test dataset

Method	PSNR (dB)/SSIM		
	$\sigma = 0.1$	$\sigma = 0.3$	$\sigma = 0.8$
Lee [4]	26.14/0.841	26.11/0.837	26.04/0.832
Frost [5]	23.9/0.747	23.90/0.746	23.89/0.745
Kuan [6]	18.66/0.591	18.67/0.591	18.66/0.590
AD [7]	38.67/0.988	35.04/0.944	32.93/0.911
NLM [9]	40.27/0.967	37.11/0.953	34.95/0.943
BM3D [12]	24.55/0.742	24.55/0.742	24.53/0.742
DnCNN [13]	45.68/0.993	39.38/0.979	37.5/0.972
US-Net (proposed)	**46.30/0.995**	**39.45/0.980**	**37.53/0.974**

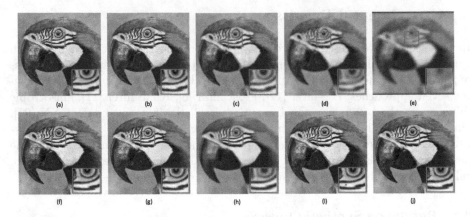

Fig. 3. Denoising results on Set12 dataset with noise level 0.3 by different methods. (a)original image (b) Noisy (38.15 dB) (c) Lee (25.92 dB) (d) Frost (23.63 dB) (e) Kuan (17.87 dB) (f) AD (36.85 dB) (g) NLM (38.86 dB) (h) BM3D (25.07 dB) (i) DnCNN (38.64 dB) (j) US-Net (40.85 dB)

Table 2. Numerical results of different methods on 1724 thyroid Ultrasound images

Method	PSNR(dB)/SSIM		
	$\sigma = 0.3$	$\sigma = 0.8$	$\sigma = 1.5$
Lee [4]	34.33/0.933	33.77/0.925	32.56/0.904
Frost [5]	29.88/0.807	29.79/0.815	29.62/0.813
Kuan [6]	23.33/0.688	23.36/0.691	23.37/0.69
AD [7]	39.54/0.977	32.20/0.879	27.64/0.729
NLM [9]	40.88/0.981	34.61/0.928	31.75/0.871
BM3D [12]	28.55/0.737	28.52/0.736	28.47/0.735
DnCNN [13]	40.83/0.980	**37.88/0.969**	34.01/0.938
US-Net (proposed)	**41.73/0.987**	37.47/**0.969**	**34.12/0.940**

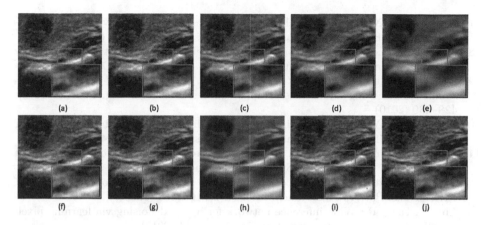

Fig. 4. Results on Ultrasound dataset with noise level 0.3 by different methods. (a) original image (b) Noisy (44.43 dB) (c) Lee (37.18 dB) (d) Frost (28.21 dB) (e) Kuan (26.12 dB) (f) AD (41.25 dB) (g) NLM (44.39 dB) (h) BM3D (29.98 dB) (i) DnCNN (40.83 dB) (j) US-Net (45.79 dB)

4 Conclusion

In this paper, we proposed a new deep convolutional neural network model for ultrasound image denoising. We also designed a new hybrid loss to make the training process faster and more stable for speckle noise removal. Experimental results on synthetic and real ultrasound images show that the our proposed model has promising performance and state-of-the-art results.

References

1. Betrouni, N., Vermandel, M., Pasquier, D., Maouche, S., Rousseau, J.: Segmentation of abdominal ultrasound images of the prostate using a priori information and an adapted noise filter. Comput. Med. Imaging Graph. **29**(1), 43–51 (2005)
2. Buades, A., Coll, B., Morel, J.M.: A non-local algorithm for image denoising. In: 2005 IEEE Computer Society Conference on Computer Vision and Pattern Recognition (CVPR 2005), vol. 2, pp. 60–65. IEEE (2005)
3. Chen, D.R., et al.: Classification of breast ultrasound images using fractal feature. Clin. Imaging **29**(4), 235–245 (2005)
4. Coupé, P., Hellier, P., Kervrann, C., Barillot, C.: Nonlocal means-based speckle filtering for ultrasound images. IEEE Trans. Image Process. **18**(10), 2221–2229 (2009)
5. Dabov, K., Foi, A., Katkovnik, V., Egiazarian, K.: Image denoising by sparse 3-D transform-domain collaborative filtering. IEEE Trans. Image Process. **16**(8), 2080–2095 (2007). https://doi.org/10.1109/TIP.2007.901238
6. Frost, V.S., Stiles, J.A., Shanmugan, K.S., Holtzman, J.C.: A model for radar images and its application to adaptive digital filtering of multiplicative noise. IEEE Trans. Pattern Anal. Mach. Intell. **4**(2), 157–166 (1982)

7. Huang, L.L., Xiao, L., Wei, Z.H.: Multiplicative noise removal via a novel variational model. EURASIP J. Image Video Process. **2010**(1), 1–16 (2010)
8. Ioffe, S., Szegedy, C.: Batch normalization: accelerating deep network training by reducing internal covariate shift. arXiv preprint arXiv:1502.03167 (2015)
9. Khare, A., Khare, M., Jeong, Y., Kim, H., Jeon, M.: Despeckling of medical ultrasound images using daubechies complex wavelet transform. Sig. Process. **90**(2), 428–439 (2010)
10. Krissian, K., Westin, C.F., Kikinis, R., Vosburgh, K.G.: Oriented speckle reducing anisotropic diffusion. IEEE Trans. Image Process. **16**(5), 1412–1424 (2007)
11. Kuan, D., Sawchuk, A., Strand, T., Chavel, P.: Adaptive restoration of images with speckle. IEEE Trans. Acoust. Speech Signal Process. **35**(3), 373–383 (1987)
12. Lee, J.S.: Digital image enhancement and noise filtering by use of local statistics. IEEE Trans. Pattern Anal. Mach. Intell. **2**(2), 165–168 (1980)
13. Liu, P., Fang, R.: Wide inference network for image denoising via learning pixel-distribution prior. arXiv preprint arXiv:1707.05414 (2017)
14. Loupas, T.: Digital image processing for noise reduction in medical ultrasonics. University of Edinburgh (1988)
15. Martin, D., Fowlkes, C., Tal, D., Malik, J., et al.: A database of human segmented natural images and its application to evaluating segmentation algorithms and measuring ecological statistics. In: ICCV, Vancouver (2001)
16. Noh, H., Hong, S., Han, B.: Learning deconvolution network for semantic segmentation. In: Proceedings of the IEEE International Conference on Computer Vision, pp. 1520–1528 (2015)
17. Wang, S., Huang, T.Z., Zhao, X.L., Mei, J.J., Huang, J.: Speckle noise removal in ultrasound images by first-and second-order total variation. Numer. Algorithms **78**(2), 513–533 (2018)
18. Wang, T., Qin, Z., Zhu, M.: An ELU network with total variation for image denoising. In: Liu, D., Xie, S., Li, Y., Zhao, D., El-Alfy, E.S. (eds.) Neural Information Processing, pp. 227–237. Springer, Cham (2017). https://doi.org/10.1007/978-3-319-70090-8_24
19. Wu, D., Kim, K.S., Fakhri, G.E., Li, Q.: A cascaded convolutional nerual network for x-ray low-dose CT image denoising. CoRR abs/1705.04267 (2017). http://arxiv.org/abs/1705.04267
20. Yu, Y., Acton, S.T.: Speckle reducing anisotropic diffusion. IEEE Trans. Image Process. **11**(11), 1260–1270 (2002)
21. Zhang, K., Zuo, W., Chen, Y., Meng, D., Zhang, L.: Beyond a gaussian denoiser: residual learning of deep CNN for image denoising. IEEE Trans. Image Process. **26**(7), 3142–3155 (2017). https://doi.org/10.1109/TIP.2017.2662206
22. Zhang, K., Zuo, W., Gu, S., Zhang, L.: Learning deep CNN denoiser prior for image restoration. In: Proceedings of the IEEE Conference on Computer Vision and Pattern Recognition, pp. 3929–3938 (2017)
23. Zhang, K., Zuo, W., Zhang, L.: FFDNet: toward a fast and flexible solution for CNN-based image denoising. IEEE Trans. Image Process. **27**(9), 4608–4622 (2018)
24. Zhao, H., Gallo, O., Frosio, I., Kautz, J.: Loss functions for image restoration with neural networks. IEEE Trans. Comput. Imaging **3**(1), 47–57 (2017)

Automatic Sinus Surgery Skill Assessment Based on Instrument Segmentation and Tracking in Endoscopic Video

Shan Lin[1]([✉]), Fangbo Qin[2], Randall A. Bly[1], Kris S. Moe[1], and Blake Hannaford[1]

[1] University of Washington, Seattle, WA 98195, USA
{shanl3,Randbly,krismoe,blake}@uw.edu

[2] Institute of Automation, Chinese Academy of Sciences, Beijing 100190, China
qinfangbo2013@ia.ac.cn

Abstract. Current surgical skill assessment mainly relies on evaluations by senior surgeons, a tedious process influenced by subjectivity. The contradiction between a growing number of surgical techniques and the duty-hour limits for residents leads to an increasing need for effective surgical skill assessment. In this paper, we explore an automatic surgical skill assessment method by tracking and analyzing the surgery trajectories in a new dataset of endoscopic cadaveric trans-nasal sinus surgery videos. The tracking is performed by combining the deep convolutional neural network based segmentation and the dense optical flow algorithm. Then the heat maps and motion metrics of the tip trajectories are extracted and analyzed. The proposed method has been tested in 10 endoscopic videos of sinus surgery performed by 4 expert and 5 novice surgeons, showing the potential for the automatic surgical skill assessment.

Keywords: Instrument segmentation · Surgical skill assessment · Sinus surgery

1 Introduction

The current mainstream surgical training paradigm, which is an apprenticeship model proposed by Dr. Halsted, has been used for more than a century [3]. Residents assist or perform surgeries under the supervision of senior surgeons, who also rate the residents' surgical skill [13]. This method suffers from subjectivity and is time-consuming. To standardize the evaluation process, rating criteria such as Objective Structured Assessment of Technical Skill (OSATS) has been proposed, but these rating methods are still tedious and have a certain degree of subjectivity [1]. In addition, due to the advent of new surgical techniques and regulations, residents need to acquire more skills but their training time is restricted by mandated duty-hour limits [13]. Therefore, more automated and efficient surgical skill assessment methods are needed.

© Springer Nature Switzerland AG 2020
Q. Li et al. (Eds.): MMMI 2019, LNCS 11977, pp. 93–100, 2020.
https://doi.org/10.1007/978-3-030-37969-8_12

Motion tracking is a significant component of the skill assessment systems [13]. Commonly used tracking systems, including optical, acoustic, electromagnetic and RFID systems [2], may change the surgery experience and constrain the surgeons' movements [13]. In addition, they are generally expensive, bulky and may not be sterilizable [13]. Surgical robotic systems provide rich motion data, including kinematic, dynamic, visual and haptic information, that have been applied for skill assessment [6,9]. Recently, simulation-based training systems have been studied for objective assessment [8]. However, these systems still have many differences from the real surgery environment.

As an alternative approach, video-based tracking is cost-efficient, noninvasive and widely applicable. Early studies used color or shape markers, but this method suffers from occlusion, may require modification to the instruments and thus can interfere with the surgical workflow [2]. To avoid these challenges, detection and tracking without visual markers has become popular [2]. However, marker-less methods need to address many challenging conditions in the surgery scene, including co-axial illumination conditions, occlusion by tissue or instruments and blur from motion [2]. Furthermore, most existing skill assessment methods are based on 3D information [6,9], but calculating 3D information from the stereoscopic video is difficult. Features to be matched in the endoscopic video are very limited. In some cases, only monocular vision is available [2], making the problem even more complex.

Considering that many classic surgical skill assessments are conducted by watching videos, our hypothesis is that 2D data contain information of surgical skill comparable to 3D data. However, there are few studies of surgical skill assessment based only on video data. Jin *et al.* [7] tracked instrument tips in real laparoscopic surgery videos using deep learning and then analyzed metrics for skill assessment. Oropesa *et al.* [12] analyzed motion metrics based on the 3D trajectories calculated from endoscopic videos in a box trainer setup.

In this study, we build a dataset including 10 endoscopic sinus surgery videos and conduct the first video-based study of endoscopic sinus surgery, showing the potential for cost-effective and standard objective assessment. We compare three advanced neural networks for instrument segmentation and track the instrument tip based on the segmentations by integrating the dense optical flow algorithm [5] and a geometric method. Metrics based on 2D tracking are compared with metrics extracted from 3D ground truth collected by tracking systems.

2 Methods

To analyze the movements of instruments, we extracted 2D trajectories of the instrument tip from monocular frames by combining image segmentation and tracking techniques. We chose this strategy because there is no significant marker or sharp corner on the tip (Fig. 1), it is not suitable to use keypoint detector or object detector to localize the tip.

2.1 Deep Neural Networks for Instrument Segmentation

We implemented segmentation with three convolutional neural network based models, DeepLabV3, TernausNet16 and the baseline model FCN8s [4,10,14]. We used the ResNet-50 pretrained on ImageNet for feature extraction in DeepLabV3. The atrous rates of the atrous spatial pyramid pooling module were 2 and 4. We used VGG-16 pretrained on ImageNet to extract feature in both TernausNet16 and FCN8s.

After comparison, we adopted the mean squared error loss instead of the cross-entropy loss. The dropout layer with keep probability of 0.5 was embedded after the output layer of ResNet-50 and VGG-16. For data augmentation, we randomly flipped and rotated the training images, and randomly jittered their brightness, hue, and contrast. We trained the network with 20 epochs through Adam optimizer using an initial learning rate of 0.001 with a decay rate of 0.5 and step of 5 epoch. See Sect. 3.1 for details of training and test dataset.

2.2 Tracking

To analyze the movements of the instrument and endoscope, we generated the segmentations of all frames in the videos and implemented instrument tip tracking by combining a geometric method with dense optical flow algorithm using the following steps:

1. Post-processing: In our dataset (Sect. 3.1) there is at most one instrument in each frame. Therefore, the instrument is the largest blob with an area greater than a threshold determined empirically.
2. Detect the tip based on a geometric method (Fig. 2): Intuitively, the tip is the farthest point from the center point of the instrument contour point along the edge of the endoscope video. To guarantee the accuracy of detection, only the contour points near the line connecting the center of edge and center of the classified region are considered.
3. Track instrument using dense optical flow algorithm: The optical flow of each pixel in the frame is calculated through the Gunner Farneback algorithm [5]. The optical flow vector of the instrument is then calculated as the average optical flow within the segmented instrument region.
4. Estimate tip position: Using a small region around the tip detected in the last frame as a template, we calculate the cross-correlations between the template and two potential positions calculated in step 2 and 3. The tip position will then be the weighted sum of these two candidates based on the cross-correlations.
5. Remove sudden change on the trajectory: A 11th-order low-pass Butterworth filter with the cutoff frequency of 0.2 was applied to smooth the trajectories.

In addition, we estimated the velocity of background based on dense optical flow. We calculated the average magnitude of the optical flow vectors in the background region. Because the tissue may deform when contacting the instrument, the pixels with a distance of 20 pixels to the instrument region are not considered in background flow estimation.

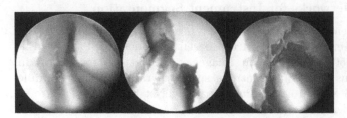

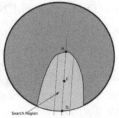

Fig. 1. Example frames in the dataset.

Fig. 2. Schematic diagram of the geometric method.

2.3 Metrics for Surgical Skill Assessment

We plotted the heat map of each instrument tip trajectory and calculated the distances between the centers of heat maps and the centers of endoscope videos. From both 2D and 3D trajectories, we extracted the total operative time, path length, average speed and acceleration of instrument. In addition, economy of area was calculated as the maximum area covered by the instrument in the video divided by the 2D path length, and economy of volume was the maximum volume covered by the instrument in the space divided by the 3D path length. These two metrics (introduced by [12]) reveal usage efficiency of the working space.

3 Experimental Setup and Results

3.1 Dataset

Ten endoscopic sinus surgeries conducted by 4 senior surgeons and 5 residents on the left and right sides of 5 cadavers were recorded at Harborview Medical Center using a Stryker 1088 HD camera system and the Karl Storz Hopkins Ø4mm 0° endoscope with a frame rate of 29.97 and a resolution of 320 × 240. Simultaneously, the location of the instrument and endoscope were recorded by the Medtronic Stealth Station S7 surgical navigation system. In one of 5 operations by expert surgeon, the video #4, the video data was truncated after maxillary antrostomy and does not include ethmoidectomy, as the others do.

We extracted a total of 3871 frames from 10 videos (Fig. 1) and manually labeled the bounding polygons of instruments. The dataset is separated into a training set of 2375 frames which corresponds to 7 videos and a test set of 1496 frames which corresponds to the remaining 3 videos. The challenging conditions of this dataset include specular reflections, low resolution, and occlusions by anatomy.

3.2 Segmentation and Tracking Results

The segmentation performance is evaluated by mean Dice coefficient, mean Intersection over Union (IoU) and training time (Table 1). DeepLabv3 surpasses

FCN-8s and TernausNet-16 with a mean Dice of 0.935 and mean IoU of 0.901. Moreover, while achieving the best performance, the training time for DeepLabv3 is more than four times less than FCN-8s and two times less than TernausNet-16.

Figure 3 shows the heat maps of the trajectories and Fig. 5b is the boxplot of the distances between the centers of heat maps and the centers of endoscope videos. Figure 4a presents the average speed of background in each video.

Table 1. Segmentation results

Network	Mean Dice	Mean IoU	Training time (ms)
FCN-8s	0.909	0.869	81.2
TernausNet-16	0.903	0.862	40.5
DeepLabv3	**0.935**	**0.901**	**17.1**

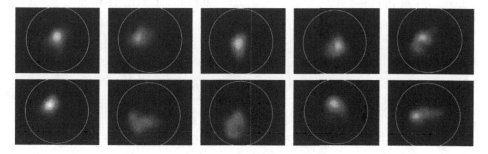

Fig. 3. Heat maps of surgery trajectories (endoscope screen coordinates). Brighter points correspond to a higher frequency of the instrument tip location within endoscopic view during surgery. Maps in the top row are from senior surgeons and maps in the bottom row are from residents.

3.3   Evaluation of Motion Metrics

Figures 5 and 4 present the extracted metrics. T-test has been applied to determine if there is a significant difference between the senior surgeons and residents based on each metric (Table 2). Two 2D metrics, path length and economy of area, have a p-value less than 0.05, while all 3D metrics have p-value greater than 0.05. However, all metrics cannot be well distinguished using multiple testing correction methods including Bonferroni correction and correction via false discovery rate (FDR) calculated using the Benjamini-Hochberg procedure [11]. Considering the dataset with five samples in each group may be too small, we performed a statistical power analysis on each metric to determine the minimum sample size required assuming the validity of the sample variance and effect size (Table 2).

Table 2. T-test results of each metric.

Metrics	t-statistic	p-value	Sample size w/90% power
Total operative time	−2.115	0.067	7
2D metrics			
Distance of heat map center to video center	−2.001	0.080	7
Path length	−3.381	**0.010**	5
Average speed	0.721	0.491	43
Average acceleration	−0.907	0.391	28
Economy of area	3.060	**0.016**	5
3D metrics			
Path length	−0.724	0.489	90
Average speed	1.761	0.116	10
Average acceleration	1.629	0.142	7
Economy of volume	0.638	0.541	83

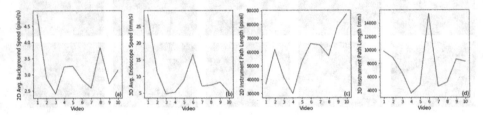

Fig. 4. (a) Average background speed in video; (b) average endoscope speed in 3D space; (c) instrument path length in video; (d) instrument path length in 3D space.

4   Discussion

While our metrics did not reveal a statistically significant difference between experienced and novice surgeons according to the Bonferroni correction results, several interesting observations can be made about the measurements arising from our new dataset. Figures 3 and 5b indicate that the expert surgeons generally maintain the instrument tip in a smaller region closer to the center of the endoscope video.

The complexity of the 10 cadaver surgeries were not strictly the same, so we can't make a conclusion in terms of the total operative time and path length. However, if we compare Fig. 4c and d (video # 4 is not considered), video # 1 (a highly experienced surgeon) has longer 3D path length than other videos except video # 6, while has the shortest 2D path length. Additionally, according to Fig. 4a, the average background speed in video # 1 is much higher than other videos. This indicates that the surgeon performed the operation effectively and

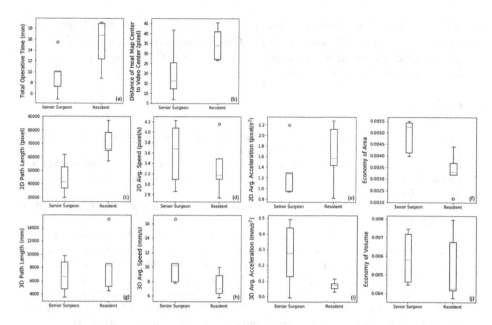

Fig. 5. (a) Total operative time; (b–f) 2D metrics: distance of heat map center to video center, path length, average speed, average acceleration, economy of area; (g–j) 3D metrics: path length, average speed, average acceleration, economy of volume.

moves the endoscope faster, and at the same time maintained a smaller relative distance between instrument and endoscope.

The 2D average speed and acceleration of the instrument do not present significant differences between the two groups. This can be explained by two causes of fast movement. Experts usually move the endoscope faster and this restricts the instrument speed to reach a very small value. For novices, it is challenging to keep the instrument stable in the endoscope video. According to power analysis, more samples are needed for more meaningful analysis.

One limitation of this work is that it does not quantitatively evaluate the 2D instrument tracking results. In the future, we will implement camera calibration and project the 3D ground truth back to the 2D space for verification. Another potential solution is to conduct 3D reconstruction based on the shape of the instrument and the calibration results, then compare the reconstruction results with the ground truth.

5   Conclusion

In summary, we introduce a new endoscopic sinus surgery dataset, propose and test a video-based surgical skill assessment method on the dataset, serving as the foundation toward automatic objective skill analysis. However, in the future study, more data is needed to improve the statistical power. Moreover, the final goal of our system is to provide instructions that help improve skills beyond statistical dexterity analysis. We anticipate this will require integration of surgery movement information and anatomy information (e.g. CT scans) to understand the intent and quality of a movement.

References

1. Aggarwal, R., Moorthy, K., Darzi, A.: Laparoscopic skills training and assessment. Br. J. Surg. **91**(12), 1549–1558 (2004)
2. Bouget, D., Allan, M., Stoyanov, D., Jannin, P.: Vision-based and marker-less surgical tool detection and tracking: a review of the literature. Med. Image Anal. **35**, 633–654 (2017)
3. Carter, B.N.: The fruition of halsted's concept of surgical training. Surgery **32**(3), 518–527 (1952)
4. Chen, L.-C., Zhu, Y., Papandreou, G., Schroff, F., Adam, H.: Encoder-decoder with atrous separable convolution for semantic image segmentation. In: Ferrari, V., Hebert, M., Sminchisescu, C., Weiss, Y. (eds.) ECCV 2018. LNCS, vol. 11211, pp. 833–851. Springer, Cham (2018). https://doi.org/10.1007/978-3-030-01234-2_49
5. Farnebäck, G.: Two-frame motion estimation based on polynomial expansion. In: Bigun, J., Gustavsson, T. (eds.) SCIA 2003. LNCS, vol. 2749, pp. 363–370. Springer, Heidelberg (2003). https://doi.org/10.1007/3-540-45103-X_50
6. Ismail Fawaz, H., Forestier, G., Weber, J., Idoumghar, L., Muller, P.-A.: Evaluating surgical skills from kinematic data using convolutional neural networks. In: Frangi, A.F., Schnabel, J.A., Davatzikos, C., Alberola-López, C., Fichtinger, G. (eds.) MICCAI 2018. LNCS, vol. 11073, pp. 214–221. Springer, Cham (2018). https://doi.org/10.1007/978-3-030-00937-3_25
7. Jin, A., et al.: Tool detection and operative skill assessment in surgical videos using region-based convolutional neural networks. In: 2018 IEEE Winter Conference on Applications of Computer Vision (WACV), pp. 691–699. IEEE (2018)
8. Lahanas, V., Georgiou, E., Loukas, C.: Surgical simulation training systems: box trainers, virtual reality and augmented reality simulators. Int. J. Adv. Robot. Autom. **1**(2), 1–9 (2016)
9. Liu, D., Jiang, T.: Deep reinforcement learning for surgical gesture segmentation and classification. In: Frangi, A.F., Schnabel, J.A., Davatzikos, C., Alberola-López, C., Fichtinger, G. (eds.) MICCAI 2018. LNCS, vol. 11073, pp. 247–255. Springer, Cham (2018). https://doi.org/10.1007/978-3-030-00937-3_29
10. Long, J., Shelhamer, E., Darrell, T.: Fully convolutional networks for semantic segmentation. In: Proceedings of the IEEE Conference on Computer Vision and Pattern Recognition, pp. 3431–3440 (2015)
11. Noble, W.S.: How does multiple testing correction work? Nat. Biotechnol. **27**(12), 1135 (2009)
12. Oropesa, I., et al.: EVA: laparoscopic instrument tracking based on endoscopic video analysis for psychomotor skills assessment. Surg. Endosc. **27**(3), 1029–1039 (2013)
13. Reiley, C.E., Lin, H.C., Yuh, D.D., Hager, G.D.: Review of methods for objective surgical skill evaluation. Surg. Endosc. **25**(2), 356–366 (2011)
14. Shvets, A.A., Rakhlin, A., Kalinin, A.A., Iglovikov, V.I.: Automatic instrument segmentation in robot-assisted surgery using deep learning. In: 2018 17th IEEE International Conference on Machine Learning and Applications (ICMLA), pp. 624–628. IEEE (2018)

U-Net Training with Instance-Layer Normalization

Xiao-Yun Zhou[1]([✉]), Peichao Li[1], Zhao-Yang Wang[1], and Guang-Zhong Yang[1,2]

[1] The Hamlyn Centre for Robotic Surgery, Imperial College London, London, UK
xiaoyun.zhou14@imperial.ac.uk
[2] Institute of Medical Robotics, Shanghai Jiao Tong University, Shanghai, China

Abstract. Normalization layers are essential in a Deep Convolutional Neural Network (DCNN). Various normalization methods have been proposed. The statistics used to normalize the feature maps can be computed at batch, channel, or instance level. However, in most of existing methods, the normalization for each layer is fixed. Batch-Instance Normalization (BIN) is one of the first proposed methods that combines two different normalization methods and achieve diverse normalization for different layers. However, two potential issues exist in BIN: first, the Clip function is not differentiable at input values of 0 and 1; second, the combined feature map is not with a normalized distribution which is harmful for signal propagation in DCNN. In this paper, an Instance-Layer Normalization (ILN) layer is proposed by using the Sigmoid function for the feature map combination, and cascading group normalization. The performance of ILN is validated on image segmentation of the Right Ventricle (RV) and Left Ventricle (LV) using U-Net as the network architecture. The results show that the proposed ILN outperforms previous traditional and popular normalization methods with noticeable accuracy improvements for most validations, supporting the effectiveness of the proposed ILN.

Keywords: Instance-Layer Normalization · Deep Convolutional Neural Network · U-Net · Biomedical image segmentation

1 Introduction

Biomedical image segmentation is a fundamental step in medical image analysis, i.e., 3D shape instantiation for organs [17] and prosthesis [14,15]. Most current popular methods are based on Deep Convolutional Neural Network (DCNN) which train multiple non-linear modules for feature extraction and pixel classification with both higher automation and performance. One fundamental component in DCNN is the normalization layer. Initially, one of the main motivations for normalization was to alleviate the internal covariate shift where layers' input distribution changes [4]. However, recent work considers the use of normalization layer is beneficial, because it increases the robustness of the networks to fluctuation associated with random initialization [2], or it achieves smoother

© Springer Nature Switzerland AG 2020
Q. Li et al. (Eds.): MMMI 2019, LNCS 11977, pp. 101–108, 2020.
https://doi.org/10.1007/978-3-030-37969-8_13

optimization landscape [9]. In this paper, we keep this motivation question open and focus on normalization strategies.

For a feature map with dimension of (N, H, W, C), where N is the batch size, H is the feature height, W is the feature width, C is the feature channel, Batch Normalization (BN) [3,4] was the first proposed normalization method which calculated the mean and variance of a feature map along the (N, H, W) dimension, then re-scaled and re-translated the normalized feature map with additional trainable parameters to preserve the DCNN representation ability. Instance Normalization (IN) [10] which calculated the mean and variance along the (H, W) dimension was proposed for fast stylization. Layer Normalization (LN) [1] which calculated the mean and variance along the (H, W, C) dimension was proposed for recurrent networks. Group Normalization (GN) [12] calculated the mean and variance along the (H, W) and multiple-channels dimension and was validated on image classification and instance segmentation. A review of these four normalization methods for training U-Net for medical image segmentation could be found in [16]. Weight normalization [8,13] based on re-parameterization on weights was used in recurrent models and reinforcement learning. Batch Kalman normalization estimated the mean and variance considering all preceding layers [11].

Recently, Nam *et al.* proposed Batch-Instance Normalization [5] (BIN), which combined BN and IN with a trainable parameter. However, two risks potentially exist: (1) the trainable parameter was restricted in the range of [0, 1] with Clip function which is not differentiable at input values of 0 and 1; (2) the combined feature map was no longer with a normalized distribution, which is harmful for signal propagation in DCNN. In this paper, Instance-Layer Normalization (ILN) is proposed to combine IN and LN: (1) Sigmoid is used to solve the non-differentiable characteristic of Clip function at input values of 0 and 1; (2) an additional GN16 - GN with a group number of 16 is added after the combined feature map to ensure a normalized distribution of the combined feature map. A widely-applied and popular network architecture - U-Net [7] is used as the network to validate the proposed ILN on the Right Ventricle (RV) and Left Ventricle (LV) image segmentation. The proposed ILN outperforms existing normalization methods with noticeable accuracy improvements in most validations in terms of the Dice Similarity Coefficient (DSC).

2   Methodology

2.1   Instance-Layer Normalization

Instance Normalization. With a feature map $\mathbf{F}$ of dimension (N, H, W, C), IN calculates the mean and variance of $\mathbf{F}$ as:

$$\mu_{n,c} = \frac{1}{H \times W} \sum_{h=1}^{H} \sum_{w=1}^{W} f_{n,h,w,c}; \quad \delta_{n,c}^2 = \frac{1}{H \times W} \sum_{h=1}^{H} \sum_{w=1}^{W} (f_{n,h,w,c} - \mu_{n,c})^2 \quad (1)$$

Then, the feature map is normalized as $\hat{\mathbf{F}}^{I}$:

$$\hat{f}^{I}_{n,h,w,c} = \frac{f_{n,h,w,c} - \mu_{n,c}}{\sqrt{\delta^{2}_{n,c} + \epsilon}} \tag{2}$$

where ϵ is a small value added for division stability. For the same feature map $\mathbf{F}$, LN calculates the mean and variance as:

$$\mu_{n} = \frac{1}{H \times W \times C} \sum_{h=1}^{H} \sum_{w=1}^{W} \sum_{c=1}^{C} f_{n,h,w,c}; \delta^{2}_{n} = \frac{1}{H \times W \times C} \sum_{h=1}^{H} \sum_{w=1}^{W} \sum_{c=1}^{C} (f_{n,h,w,c} - \mu_{n})^{2} \tag{3}$$

where $\mathbf{F}$ is normalized in a similar way of Eq. (2) to $\hat{\mathbf{F}}^{L}$. A trainable parameter ρ is added to combine $\hat{\mathbf{F}}^{I}$ and $\hat{\mathbf{F}}^{L}$. In the original BIN [5], ρ was clipped to be in the range of $[0, 1]$ with a Clip function, as shown in Fig. 1.

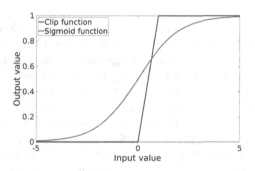

Fig. 1. The curves of Clip and Sigmoid function.

However, Clip function is not differentiable at input values of 0 and 1. In this paper, Sigmoid function $Sigmoid(x) = 1/(e^{-x} + 1)$ which is differentiable everywhere is applied to solve this potential issue:

$$\hat{\mathbf{F}}^{IL} = Sigmoid(\rho) \cdot \hat{\mathbf{F}}^{I} + (1 - Sigmoid(\rho)) \cdot \hat{\mathbf{F}}^{L} \tag{4}$$

An additional potential issue in the original BIN is that the combined $\hat{\mathbf{F}}^{IL}$ is no longer with a mean of 0 and a variance of 1, this non-normalized distribution may be harmful for signal propagation in DCNN. In this paper, we solve this issue with applying an additional GN16 on the combined $\hat{\mathbf{F}}^{IL}$:

$$\mu_{n,g} = \frac{1}{H \times W \times M} \sum_{h=1}^{H} \sum_{w=1}^{W} \sum_{m=(g-1) \cdot M+1}^{g \cdot M} \hat{f}^{IL}_{n,h,w,m}, M = C//16 \tag{5}$$

$$\delta^{2}_{n,g} = \frac{1}{H \times W \times M} \sum_{h=1}^{H} \sum_{w=1}^{W} \sum_{m=(g-1) \cdot M+1}^{g \cdot M} (\hat{f}^{IL}_{n,h,w,m} - \mu_{n,g})^{2}, M = C//16 \tag{6}$$

where M is the channel number in each feature group, $//$ is exact division, $g \in [1, 16]$. The feature map is normalized in a similar way of Eq. (2) as $\hat{\mathbf{F}}^{ILN}$. Following BN [4], additional parameters γ and β are added to preserve the DCNN representation ability $f'^{ILN}_{n,h,w,c} = \gamma_c \hat{f}^{ILN}_{n,h,w,c} + \beta_c$.

2.2 Experimental Setup

Network Architecture. A widely adopted network architecture in medical image segmentation, called U-Net [7], was used as the fundamental network framework with four max-pooling layers. The start feature channel number is 16. The normalization layer was added between the convolutional and Relu layer. Cross-entropy was used as the loss function. Momentum Stochastic Gradient Descent (SGD) was used as the optimizer with the momentum set as 0.9. Weights were initialized with a truncated normal distribution with the *stddev* as $2/(3^2 \times C)$, where C is the channel number. Biases were initialized as 0.1. ρ was initialized as 0.5.

Data Collections. 6082 RV images [17], scanned with a 1.5T Magnetic Resonance Imaging (MRI) machine (Sonata, Siemens, Erlangen, Germany), with slice gap of 10 mm, pixel spacing of 1.5~2 mm, image size of 256×256, from 37 subjects mixed with Hypertrophic Cardiomyopathy (HCM) patients and asymptomatic subjects, from the atrioventricular ring to the apex were used for the validation. The ground truth was labeled by one expert with Analyze (AnalyzeDirect, Inc, Overland Park, KS, USA). $[-30° : 10° : 30°]$ rotations were applied to augment the training images. 12, 12, 13 subjects for each group were split randomly for three-fold cross validation. 805 LV images [6], from SunnyBrook MRI dataset, with subject number of 45, image size of 256×256, were used for the validation as well. $[-60° : 2° : 60°]$ rotations were applied to augment the training images. 15 subjects for each group were split randomly for three-fold cross validation.

Implementation. As the proposed ILN needs to manipulate intermediate feature maps, the U-Net framework was implemented with low-level Tensorflow functions - tf.nn. In this paper, to ensure a fair comparison, all normalization methods were re-implemented into the same framework as the ILN implementation instead of using the available high-level Tensorflow Application Programming Interface (API) exists for some normalization methods in Tensorflow library, such as those used in [16].

Experiments. Following [16] and [18], two epochs were trained for each experiment with dividing the learning rate by 5 at the second epoch. Five initial learning rates (1.5, 1.0, 0.5, 0.1, 0.05) were tested for each experiment and the best value was selected to be shown. DSC was used as the evaluation metric.

3 Result

To prove the advantage of using the Sigmoid function over the Clip function (in original BIN [5]), three comparison experiments were set up: (1) using Clip function with one trainable parameter $Clip(\rho)_0^1$ for IN feature map while the parameter for LN feature map is $1 - Clip(\rho)_0^1$; (2) using Sigmoid function with one trainable parameter $Sigmoid(\rho)$ for IN feature map while the parameter for LN feature map is $1 - Sigmoid(\rho)$; (3) using Softmax function with two trainable parameters $Softmax(\rho_1, \rho_2)$ for IN and LN feature map respectively. Comparison results are shown in Sect. 3.1.

To prove the advantage of adding GN16 after the combined feature map, two comparison experiments with or without GN16 are conducted. Results are shown in Sect. 3.2. Eight randomly-selected segmentation examples are shown in Sect. 3.3 for intuitive illustrations. As GN16 performed similarly to IN [16], no normalization, IN, LN, GN4 are chosen as the baseline to validate the performance of the proposed ILN, as presented in details in Sect. 3.4. The training curves of ρ at eight randomly-selected layers are shown in Sect. 3.5. In this paper, RV-1 refers to the cross validation that uses the first group of RV data as testing while uses the second and third group of RV data as training. Similar fashions were applied as the notations of the experiments on the RV-2, RV-3, LV-1, LV-2, and LV-3.

3.1 Sigmoid vs. Clip vs. Softmax Function

The mean $\pm$ std segmentation DSCs of using Clip, Sigmoid and Softmax function to combine the IN and LN feature map are shown in Table 1. We can see that Sigmoid function achieves the highest DSC for most cross validations, except RV-1 experiment, which proves the effectiveness of the proposed method in this paper - replacing the Clip function in original BIN [5] with Sigmoid function.

Table 1. Mean $\pm$ std segmentation DSCs of using Clip, Sigmoid and Softmax function to combine the feature map of IN and LN, highest DSCs are in bold.

Method	RV-1	RV-2	RV-3	LV-1	LV-2	LV-3
Clip	**0.702 ± 0.295**	0.707 ± 0.299	0.666 ± 0.319	0.900 ± 0.099	0.864 ± 0.184	0.804 ± 0.246
Sigmoid	0.692 ± 0.304	**0.724 ± 0.284**	**0.675 ± 0.301**	**0.903 ± 0.118**	**0.888 ± 0.135**	**0.828 ± 0.189**
Softmax	0.688 ± 0.290	0.720 ± 0.279	0.664 ± 0.323	0.895 ± 0.151	0.866 ± 0.153	0.827 ± 0.228

3.2 With or Without GN16

The mean $\pm$ std segmentation DSCs of adding or not adding GN16 after the combined feature map of IN and LN are shown in Table 2. We can see that, the method with adding GN16 achieves the highest DSC for most cross validations, except LV-3 experiment. This result proves the effectiveness of adding GN16 after the combined feature map and also proves the importance of maintaining the normalized distribution of feature maps.

Table 2. Mean ± std segmentation DSCs of adding or not adding GN16 after the combined feature map of IN and LN, highest DSCs are in bold.

Method	RV-1	RV-2	RV-3	LV-1	LV-2	LV-3
No	0.692 ± 0.304	0.724 ± 0.284	0.675 ± 0.301	0.903 ± 0.118	0.888 ± 0.135	**0.828 ± 0.189**
Yes	**0.714 ± 0.290**	**0.737 ± 0.267**	**0.680 ± 0.305**	**0.919 ± 0.098**	**0.893 ± 0.127**	0.827 ± 0.211

3.3   Segmentation Examples

Eight segmentation examples were selected randomly from the RV and LV data to show the segmentation details in Fig. 2. For most cases, both the RV and LV are segmented properly. However, for cases near the RV apex, i.e., the forth figure in the first row, the segmentation quality is worse. This might be due to the tissue adhesion and the small size of RV.

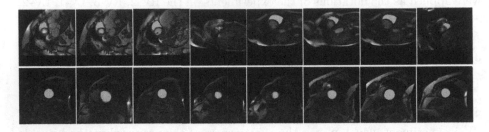

Fig. 2. Eight examples were selected randomly from the RV and LV segmentation results, where red indicates the ground truth, green indicates the segmentation result, and yellow indicates the true positives of the prediction. (Color figure online)

3.4   Comparison to Other Methods

The mean ± std segmentation DSCs of using no normalization, IN, LN, GN4, and the proposed ILN with the U-Net framework are shown in Table 3. We can see that, except the LV-3 experiment, the proposed ILN outperforms all other traditional methods with considerable accuracy improvements. This result proves the effectiveness of the proposed ILN in medical image segmentation.

3.5   Training Curves of ρ

The ρ training curves of eight layers were selected randomly from LV-1 experiment to be shown in Fig. 3. We can see that ρ was trained to be different values and the proposed ILN achieved diverse normalization at different layers. As the ground truth of ρ is not known and it is impossible to judge the curve correctness, a comparison regarding the ρ training curves of ILN and BIN is not illustrated.

Table 3. Mean ± std segmentation DSCs of using no normalization, IN, LN, GN4, and the proposed ILN with the U-Net framework, highest DSCs are in bold.

Method	RV-1	RV-2	RV-3	LV-1	LV-2	LV-3
None	0.688 ± 0.296	0.678 ± 0.318	0.661 ± 0.323	0.899 ± 0.134	0.872 ± 0.167	0.784 ± 0.280
IN	0.709 ± 0.266	0.715 ± 0.278	0.655 ± 0.327	0.905 ± 0.114	0.876 ± 0.131	$\mathbf{0.836 \pm 0.207}$
LN	0.702 ± 0.287	0.718 ± 0.270	0.662 ± 0.309	0.898 ± 0.120	0.858 ± 0.187	0.793 ± 0.262
GN4	0.679 ± 0.303	0.701 ± 0.291	0.671 ± 0.309	0.908 ± 0.113	0.841 ± 0.196	0.800 ± 0.255
ILN	$\mathbf{0.714 \pm 0.290}$	$\mathbf{0.737 \pm 0.267}$	$\mathbf{0.680 \pm 0.305}$	$\mathbf{0.919 \pm 0.098}$	$\mathbf{0.893 \pm 0.127}$	0.827 ± 0.211

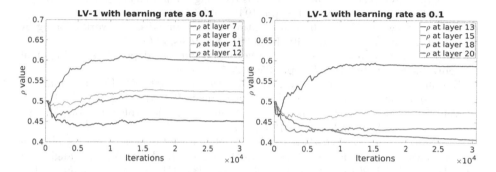

Fig. 3. The training curves of eight ρ selected randomly from the 22 layers in U-Net.

The CPU used is Intel Xeon(R) E5-1650 v4@3.60 GHz×12. The GPU used is Nvidia Titan XP. Comparing ILN to IN, the parameter number increases 22, as one parameter is added to each layer. The training time for 200 iterations increases from 34.8 s to 36.5 s due to the additional GN16 calculation.

4   Discussion

The proposed ILN strategy is generic and flexible. The three components, IN, LN and GN16 could be replaced with other normalization methods. The proposed ILN framework is validated on medical image segmentation with a U-Net framework. We believe that it could also be useful for other tasks, which needs further validation and exploration. The proposed ILN failed to achieve the highest DSC for the LV-3 experiment. It may due to that the combination of IN, LN and GN16 is not suitable for this experiment. In the future, the proposed ILN framework would be extended to combining more normalization methods.

5   Conclusion

To improve the accuracy of biomedical image segmentation based on U-net, the ILN was proposed to combine the feature map of IN and LN with an additional trainable parameter and Sigmoid function, then add GN16 after the combined feature map. Although, various normalization methods have been proposed, the noticeable accuracy improvements of the proposed ILN - almost 2% DSC proves the importance of carefully tuning the normalization strategy when training DCNNs.

References

1. Ba, J.L., Kiros, J.R., Hinton, G.E.: Layer normalization. Stat **1050**, 21 (2016)
2. Bjorck, N., Gomes, C.P., Selman, B., Weinberger, K.Q.: Understanding batch normalization. In: NeurIPS, pp. 7705–7716 (2018)
3. Ioffe, S.: Batch renormalization: towards reducing minibatch dependence in batch-normalized models. In: NeurIPS, pp. 1945–1953 (2017)
4. Ioffe, S., Szegedy, C.: Batch normalization: accelerating deep network training by reducing internal covariate shift. In: ICML, pp. 448–456 (2015)
5. Nam, H., Kim, H.E.: Batch-instance normalization for adaptively style-invariant neural networks. In: NeurIPS, pp. 2563–2572 (2018)
6. Radau, P., Lu, Y., Connelly, K., Paul, G., Dick, A., Wright, G.: Evaluation framework for algorithms segmenting short axis cardiac MRI. MIDAS J. Card. MR Left Ventricle Segmentation Challenge **49** (2009). https://ieeexplore.ieee.org/stamp/stamp.jsp?tp=&arnumber=8759179%20[12]
7. Ronneberger, O., Fischer, P., Brox, T.: U-Net: convolutional networks for biomedical image segmentation. In: Navab, N., Hornegger, J., Wells, W.M., Frangi, A.F. (eds.) MICCAI 2015. LNCS, vol. 9351, pp. 234–241. Springer, Cham (2015). https://doi.org/10.1007/978-3-319-24574-4_28
8. Salimans, T., Kingma, D.P.: Weight normalization: a simple reparameterization to accelerate training of deep neural networks. In: NeurIPS, pp. 901–909 (2016)
9. Santurkar, S., Tsipras, D., Ilyas, A., Madry, A.: How does batch normalization help optimization? In: NeurIPS, pp. 2488–2498 (2018)
10. Ulyanov, D., Vedaldi, A., Lempitsky, V.: Instance normalization: The missing ingredient for fast stylization. arXiv preprint arXiv:1607.08022 (2016)
11. Wang, G., Peng, J., Luo, P., Wang, X., Lin, L.: Batch kalman normalization: Towards training deep neural networks with micro-batches. arXiv preprint arXiv:1802.03133 (2018)
12. Wu, Y., He, K.: Group normalization. In: Ferrari, V., Hebert, M., Sminchisescu, C., Weiss, Y. (eds.) ECCV 2018. LNCS, vol. 11217, pp. 3–19. Springer, Cham (2018). https://doi.org/10.1007/978-3-030-01261-8_1
13. Xu, Y., Wang, X.: Understanding weight normalized deep neural networks with rectified linear units. In: NeurIPS, pp. 130–139 (2018)
14. Zhou, X.Y., Lin, J., Riga, C., Yang, G.Z., Lee, S.L.: Real-time 3D shape instantiation from single fluoroscopy projection for fenestrated stent graft deployment. IEEE RAL **3**(2), 1314–1321 (2018)
15. Zhou, X.Y., Riga, C., Lee, S.L., Yang, G.Z.: Towards automatic 3D shape instantiation for deployed stent grafts: 2D multiple-class and class-imbalance marker segmentation with equally-weighted focal U-Net. In: 2018 IEEE/RSJ IROS, pp. 1261–1267 (2018)
16. Zhou, X.Y., Yang, G.Z.: Normalization in training U-Net for 2D biomedical semantic segmentation. IEEE RAL **4**(2), 1792–1799 (2019)
17. Zhou, X.Y., Yang, G.Z., Lee, S.L.: A real-time and registration-free framework for dynamic shape instantiation. MedIA **44**, 86–97 (2018)
18. Zhou, X.Y., Zheng, J.Q., Yang, G.Z.: Atrous convolutional neural network (ACNN) for biomedical semantic segmentation with dimensionally lossless feature maps. arXiv preprint arXiv:1901.09203 (2019)

Author Index